Second Edition

Tessie, Quagga Mussels, and Other Lake Tahoe Myths

DRAFT

This is a work-in-progress. If errors are detected, please submit supporting documentation or links on the Contact page at: SaveTessie.org.

Steve Urie

This material is not copyrighted and is available for reprint without restriction.

Last updated: June, 2016

ISBN: 1494777045
ISBN-13: 9781494777043

Cover design by Pond Collective · Truckee, CA

For Ethan, whose boundless enthusiasm for creatures big and small is a constant inspiration.

CONTENTS

Preface 1

Introduction 7

More Threatening than Tessie 11

It Takes a Tough Species to Live at Tahoe 27

The Winds of War 47

Scientific Ambiguity or Deceit 59

Verification & Validation 75

An Ounce of Prevention 87

A Pound of Cure 97

The Elephant in the Room 115

Lake George: Bad Advice from Lake Tahoe ... 137

Donner Lake: A Case Study 149

Financial Delusions 169

TRPA: Too Important to Fail? 181

Management Paradox 195

New Directions 211

After Notes 220

Works Cited 232

Index 240

Preface

Sometime in the early 1980's, after Donner Lake's water level was lowered below the pier footings for the winter, a fingernail-size shell caught my eye as I walked along the newly exposed shoreline of the High Sierra lake. Over the next 25 years, I occasionally spotted similar shells and would add them to a water glass collection. By 2009, my glass was more than half full. The tiny clam shells were the only shells I found along the shore of the pristine lake, where except for a few species of minnows, game-fish introductions, and crayfish there are few aquatic animals.

When I read in 2009 that a $1.4 million effort by a collaboration of federal, state, and local agencies called the Lake Tahoe Asian Clam Working Group to Eradicate the Asian Clam had been formed at neighboring Lake Tahoe, I went online to see if the shells I had collected were from the non-native clams that the Tahoe Regional Planning Agency (TRPA) was experimenting to find an effective way to eradicate—they were.

A year later, Senator Dianne Feinstein observed at the Lake Tahoe Summit that "If you organized all the Asian clams currently in the lake end-to-end, it would stretch 3.5 miles long." I was amused at the mental image that a single-file, three-mile long column of dime-size clams invoked, but when I calculated that all of Tahoe's clams would fit in the bed of a pickup truck, I became less amused at the thought of spending $1.4 million to research how to kill them, and I wrote an op-ed to the local newspaper saying so. I was surprised at the column's response. One reader wrote that I had no idea of the damage Asian clams were causing, and that if aquatic invasive species (AIS) destroyed Lake Tahoe and its tourism-based economy, it was the fault of people like me.

I watched with interest as the clams were increasingly vilified, and when in October 2012 the eradication experiments escalated into a full-blown program at Tahoe's Emerald Bay, I did some research. I found that sparse colonies of the non-native clams cause no harm, and because they remove algae, bacteria, and detritus from the water they actually improve clarity and water quality. I also wanted to learn why if the clams were as fecund and destructive as researchers claimed and had been in local lakes for years, they had only recently become a threat.

I learned that after quagga mussels were found in Nevada's Lake Mead in early 2007, there was great consternation among Western environmentalists. It was the first time that the Eastern European mussels had colonized west of the Rockies. And although ecologists knew that Sierra lakes had very low calcium and couldn't support quagga, a University of Nevada, Reno (UNR) professor hypothesized that the clam shells might increase Tahoe's dissolved calcium to a level where they could survive.

As preposterous as it was that clam shells could raise Lake Tahoe's dissolved calcium levels, it seemed harmlessly prudent when TRPA facilitated voluntary inspections to help assure that boats launching in the lake weren't carrying non-native species from infested lakes. And like most I believed the press releases and thought that it was only responsible to protect Tahoe to the highest possible degree. But as claims of potential harm by Asian clams became more unreasonable and unscientific, I realized that fears of ecological damage by the clams and other aquatic nonnatives were exaggerated. And when acres of rubber matting were laid down to purge the lake bottom of Asian clams—and all other life—I thought it was time to take a close look at the science behind Tahoe's AIS prevention and control programs.

I knew that narrow shorelines, granite substrate, and cold water make Sierra lakes inhospitable to only the hardiest aquatic plants and animals, so I researched the area's water quality, and the diet, reproductive, and habitat requirements of the alien species that were said to threaten the lakes to see if they were a genuine menace. I learned from dozens of scientific sources that several aquatic animals that are non-native to North America do measureable environmental damage, a few cause economic harm, and some threaten many U.S. regions—but at Tahoe and its nearby lakes and streams, none of them could do measurable environmental damage or cause any economic harm, and almost all couldn't survive for long in Lake Tahoe's clear, cold water. That surprised me. But I interviewed a fishing guide who explained that the lack of plant cover and the scarcity of food in High Sierra lakes prevented most fish species from establishing in the lakes, and that those that did, didn't prosper in large numbers.

I reviewed dozens of journal articles, texts, and studies on aquatic invasive species and everything available on Tahoe area AIS studies and projects. And I intensively researched quagga and zebra mussels (dreissenidae), spiny water flea and New Zealand mud snails—the species TRPA says that boat inspections will keep from infesting the lake. I found that the conclusions reached by Tahoe's researchers usually differed with accepted science and that their recommendations were biased in support of prevention and control programs. I learned that there are some aquatic species that are invasive to much of North America, but that none of the aquatic animals TRPA says would harm Lake Tahoe can establish in the lake.

Pandas and black bears are genetically close relatives, but we intuitively know that bamboo-eating pandas wouldn't make it through a Tahoe winter because their diet and habitat requirements are not suited to the High Sierra. The same is true for quagga and zebra mussels, spiny water flea, and New Zealand mud snails. But because they are unfamiliar, and even sound undesirable, they are easily vilified, and we believe professors and agency managers when they claim that non-native mollusks and crustaceans could infest Sierra Nevada lakes, damage boats and infrastructure, harm habitats, and degrade water quality.

Since 1997 TRPA has had oversight responsibility for more than $100 million dollars a year that primarily flows from the states of California and Nevada and federal sources into TRPA's Environmental Improvement Program. That kind of serious money attracts the attention of those whose pay checks are underwritten by the public, and more than 50 agencies, organizations, and public universities have cozied up with TRPA to share in the federal agency's annual largesse. And because any perceived threat to Tahoe is attacked with emotional fervor, most of those agencies and organizations have joined the TRPA-led war on non-native aquatic species.

At Tahoe, the public perception is that "one quagga larva could multiply rapidly and the resulting population could destroy Tahoe's environment and economy." The truth is that even if millions of quagga larvae were dumped into the lake, they would all die within a few days in the lake's cold, low-calcium water. But the most surprising fact I learned was that the greatest peril to Tahoe's native aquatic animals is not from invasive species but from gamefish introductions and destructive control programs.

Over a seven-year period starting in 2008, TRPA oversaw a multi-million dollar Asian clam eradication effort that culminated in carpeting five acres of Emerald Bay with synthetic black mats. After two years of falling far short of their objectives, the agency acknowledged their failure, blamed it on a lack of funding, and stowed away the mats, but vowed to continue to research new methods to control Asian clams—even though the researchers who suggested and designed the program conceded that they were wrong, and that Asian clam control is impossible anywhere.

After the failure of their Asian clam control program, in 2015 TRPA shifted the emphasis of their AIS control programs to warm-water fish and non-native plants. A quarter century earlier, Eurasian watermilfoil was first found in the Tahoe Keys, and within a decade the attractive, hardy aquatic plant that is used to decorate aquariums was clogging Keys' waterways and costing the property owners $400,000 annually to mechanically remove watermilfoil from their lagoons and canals to keep them open for navigation. Apparently, TRPA learned from their programmatic failures with Asian clam control and didn't take the lead in controlling the watermilfoil in the Tahoe Keys, but still financially supported control efforts and didn't block the Key's homeowners plan to use chemical herbicides—which had been previously prohibited—in the development behind the Lake Tahoe shoreline.

TRPA's AIS prevention program is to inspect all trailered watercraft before launching for caked mud or undrained water that may contain non-native species or their larvae, and to wash and decontaminate those boats that aren't deemed to be "clean, drained, and dry" with hot water—totally unnecessary and wasteful procedures. The purpose of this book is to explain why that is true, and why controlling non-native species in Lake Tahoe is impossible.

Steve Urie

Introduction

Many ecologists claim that a downside of globalization is a destructive worldwide migration of plants and animals. Others say that the net effect of non-native species introductions improves biodiversity and the overall result is positive. The reality is that it's a pointless argument—the homogenization of the earth's species is an inevitability that can barely be slowed, much less stopped or reversed. Most introductions are benign and unnoticed, some are beneficial, and occasionally one arises that causes economic or environmental harm and is deservedly labeled as "invasive."

Seeking favorable habitats, plant and animal migrations have occurred since the first single-cell species evolved—and soon migrated. But the rapid snow-globe shakeup of the earth's flora and fauna didn't start until the fifteenth century when European explorers not only stowed foodstuffs and seeds in their ships' holds, but also microbes, larvae, insects, rodents, and countless other vermin that tagged along.

In North America it wasn't until the 1821 completion of the Erie Canal that marine animals accessed the Great Lakes from the Atlantic Ocean. Few non-native plant and animal species are well-suited for the northern lakes' harsh environment, but about 180 non-native aquatic animals have established there since the world's largest freshwater system was connected to the oceans. One that thrived is the sea lamprey.

The adaptive eel-like fish grow to three feet in length and prey on fish greater than their own weight. The blood-sucking fish played a major role in the decline of Lake Superior's mackinaw population, and although a few human lamprey attacks have been reported, lampreys prefer cold-blooded animals. Toxic chemicals help control lampreys, but no eradication method has been found. Presently, lamprey control in the Great Lakes annually costs Canada and the United States about $20 million.

Two thirds of the Great Lakes' aquatic nonnatives were introduced before the St. Lawrence Seaway opened the lakes to ocean-going freighters in 1959. Daily, dumping millions of gallons of ballast water, foreign freighters introduced thousands of aquatic plants and animals from around the world into North America, but only a small percentage were able to establish. Aquatic biologists attribute the inability of aquatic plants and animals to propagate and colonize in the Great Lakes to various environmental factors—any one of which can prevent their establishment. Ballast water usually carries marine animals that can only survive in saltwater; rocky Great Lakes' substrate is unsupportive of aquatic plants and consequently affords poor cover for fish; a broad range of predators exists; the lakes freeze and winter water temperatures drop to the low-30's; and most importantly, nutrients and food sources are meager.

Because most non-native aquatic species were unable to secure a habitat niche or were absorbed into the Great Lakes' food web without environmental stress, ecologists were relieved. The St. Lawrence Seaway seemed a relatively benign pathway for non-native species introductions, but then in 1988, zebra mussels were found in Lake Erie, and a year later quagga mussels, their Eastern European cousins, were found there too.

Impacts from the mussels were immediate, and lakeside water treatment plants, industrial facilities, and nuclear power plants were affected when mussels attached themselves to water intake infrastructure, reducing inflow, and creating maintenance issues. Because warm water temperatures favor the mussels, nuclear power plant cooling systems were the most affected. In the early 1990's, intensive research produced several inexpensive, effective mussel control methods, and since then annual costs for monitoring and control of non-native mussels in the eight Great Lakes states and Canada has leveled off at about $30 million.

Agency managers consistently say that AIS prevention is less expensive than control. That is wrong. In 2010 the Western Regional Panel [WRP] on Aquatic Nuisance Species submitted the *Quagga-Zebra Mussel Action Plan for Western U.S. Waters* to the Aquatic Nuisance Species (ANS) Task Force, a coordinating organization of 13 federal agencies. The plan called for highest priority actions to increase capacity to address invasive mussel prevention, early-detection monitoring, rapid response, control, outreach and education, and research to address the West's growing quagga and zebra mussel issues. The estimated annual cost for only the high priority items was $117 million, of which the annual cost for boat inspections alone throughout the 19 Western States was $33 million—more than the mussel control costs in the entire Great Lakes region.

Twenty years prior to the Western action plan for mussels, in 1990 the ANS Task Force implemented a national campaign to stop the spread of quagga and zebra mussels. It was known that mussel larvae were transported by rivers and streams, aquatic birds, and migrating animals, but it was logical that the mussels also leapfrogged between isolated lakes in live bait containers and compartment and bilge water in boats. Physical transport by watercraft could be controlled, and the task force initiated an information campaign promoting the importance of cleaning, draining, and drying boats between launches at different lakes.

The outreach program didn't work. Skipping few large water bodies, during the next decade the aggressive mussels extended their range throughout the Ohio River Valley, westward into the Upper Mississippi and Missouri River Valleys, up the Red River, and down the Mississippi River to the Gulf of Mexico. While the mussels relentlessly advanced west across the Plains States, it appeared that the Rocky Mountains and the Great Basin deserts beyond the mountain range would be an effective barrier to the invasive mussels' migration west of the Continental Divide, and that the West would remain free of quagga and zebra mussels. It was not to be. In early 2007, quagga were detected in Lake Mead—the Colorado River had carried quagga larvae 800 miles from the western slope of the Rocky Mountains into the southern Nevada reservoir. When the first quagga mussels west of the Rockies were confirmed, Western environmental agencies and organizations were ready for them.

Even though two decades of experience in the Great Lakes region showed that the mussels spread easily between proximate water bodies and that no prevention measure had been effective, concerted efforts were launched to prevent the mussels from being transported to other western waterbodies. The plans failed. Within six years after their discovery in southern Nevada, quagga mussels spread to rivers, aqueducts, and lakes in Arizona, Utah, New Mexico, and California—but notably, they didn't spread to waterbodies in the Pacific Northwest or the Sierra Nevada.

Lake Mead, the Colorado River reservoir created by Hoover Dam in 1935, is 500 miles from Lake Tahoe, lies at an elevation 5,000 feet lower than North America's second deepest lake, has an average year-round water temperature 30 degrees warmer than Tahoe's, and is as ecologically different from the alpine lake as the barren desert that borders Lake Mead is from Tahoe's snow-capped peaks. However, it's not the stark geological and climate differences between Lake Mead and Lake Tahoe that prevent quagga mussels that thrive in the desert reservoirs from infesting Sierra Nevada lakes—it's the mountain lakes' incredibly pure water.

— 1 —

More Threatening than Tessie

If quagga mussels "ever make it into Lake Tahoe, they will be here forever. They would transform our beautiful beaches into mazes of jagged shells and cause a stench as they decompose. Treatment could easily cost tens of millions of dollars every year."

~ League to Save Lake Tahoe website

Lake Tahoe has stirred men's imaginations since they first saw it. For generations, tall tales of a shaken Jacques Cousteau finding a graveyard of bodies perfectly preserved by the lake's frigid waters and grimly saying, "The world isn't ready for the horrors I have seen,"[1] and of Tessie, Tahoe's serpentine monster have been passed from parent to child around campfires.

On April 1, 1992 the Tessie myth became reality. A blurred grainy, black and white photo taken by *North Lake Tahoe Bonanza* photographer, Jim Grant, emblazoned the newspaper's front page. The accompanying story relayed how the U.S. Coast Guard pulled an Incline Village man from the frigid waters of Lake Tahoe after his "motor boat was swamped by what witnesses described as a 75-foot, green lizard with a long neck." Grant and his news

---

[1] Some incidental quotes in this text are unattributed; to find the source and context, Google the quote.

editor, James Robbins, thought that everyone would get a good chuckle out of the outrageous April Fools' Day story that was enhanced by Grant's picture of a toy brontosaurus anchored by a shoelace at the edge of the lake. But the newsroom was dismayed when approximately 200 readers swamped the phone lines asking for more information about the "lake monster." Most laughed after learning about their gullibility, but one woman chastised the *Bonanza's* publisher because the story had traumatized her dogs, who walked the shoreline every morning. She was told that the bigger story was that her dogs read the *Bonanza*.

In recent years grisly new tales have started to replace old Tahoe myths. Children and their parents are now told that if the smell of rotting corpses and flesh-gashing shells strewn through the sand doesn't drive them from once-spectacular beaches, slimy green strings of goo will entangle their arms and legs if they venture into the water. More alarming are stories that the decomposing corpses and green goop was caused by alien invaders who snuck into the Tahoe Basin by hiding in boat hulls, and that the invaders have such voracious appetites that Tahoe's fish will starve—again demonstrating that many trustingly believe authorities, regardless of how outrageous a story is.

Similar to the story of Cousteau finding bodies mummified by Lake Tahoe's frigid depths, there is quasi-scientific explanation behind the new tales. Quagga and zebra mussels (dreissenae) have infested hundreds of lakes from coast to coast and caused costly maintenance and control issues in some commercial water intake systems. But just as there is no record of Cousteau visiting Tahoe, there is no chance of strands of algae entangling swimmers, fish starving, or quagga mussels colonizing the lake.

Dreissenid mussels are unable to infest Tahoe because its water is cold, it is low in nutrients essential to quagga, and most critically it has an exceptionally low concentration of calcium.

Calcium is an essential mineral for all animals, but especially for mollusks that need it to build their protective shells. High Sierra streams and lakes are so low in calcium that a peer-reviewed Oregon State University study places the region in its lowest risk category for quagga and zebra mussel invasion—a fortuitous situation shared by only nine percent of the country.

How did fears of a mussel invasion at Lake Tahoe arise? When quagga were discovered in Lake Mead, it was a threatening scenario that western environmental agencies were prepared for. When zebra mussels infested the Lower Great Lakes, an uproar ignited among water facility engineers and agency managers. Utilities and industrial plants along some Great Lakes' shorelines were finding clusters of mussels in their intake pipes. Curiously, Lake Superior was an exception. The northernmost and largest of the lakes had few mussel colonies and only minor issues. It was found that relative to the four southern Great Lakes, Lake Superior has significantly lower calcium concentrations.

After low-cost control methods were found, the furor over mussels calmed and they became a back-page issue in eastern environmental newsletters. But when southern Nevada reservoirs were discovered to be teeming with quagga, the more aggressive of the two dreissenid mussels, a new tempest flared up in the West. No lake is more loved and passionately protected than Lake Tahoe, and no one wanted to take a chance on its pristine waters becoming the Lake Erie of the West. The nation's largest alpine lake became ground zero for mussel prevention, and led by the TRPA, agencies and organizations redirected their environmental priorities to protecting Lake Tahoe from invasive mollusks.

Other than saying mussels have oblong shells while clams look like the gas station sign, even biology teachers have a difficult time explaining the physiological differences between the two classes of mollusks. Along with squid, octopus, snails, slugs, and a few other less familiar classes, they all belong to the Mollusca phylum, which includes about 85,000 different species. A primary difference between mussels and clams is that mussels can attach to hard surfaces with a tough filament called a byssus while clams bury themselves in bottom sediment, using their "foot" muscle to protectively cover themselves in silt, gravel, and sand.

In the spring of 2007, most people in northern California and Nevada were focused on an over-heated economy that was rapidly cooling and wars in the Mideast and most missed the news of quagga invading Lake Mead. But the mussels' discovery in southern Nevada didn't escape TRPA or Tahoe environmentalists and academics. No one at TRPA or any other Tahoe agency or any of the local university academics was a muscologist or expert in aquatic invasive species. Instead of consulting experts from the Great Lakes region, where dozens of government and university studies on dreissenid mussels had been completed and definitive environmental thresholds had been established for quagga and zebra mussel survival, and without assembling basic data on the physical, biological, and chemical characteristics of Tahoe's water, a conference of local academics and agency representatives was hastily convened in May 2007 at Incline Village.

At the conference, it was concluded that "invasive species pose a major threat to Lake Tahoe's future and efforts against them need to be stepped up." "The issue is our collective concern for the basin from the threat of invasive species," TRPA's Julie Regan said. "Prevention strategies include boat washing stations, where owners can wash vessels before putting them in the lake." "Boats are the most serious threat," said Phil Brozak of the U.S. Army Corps of Engineers.

A month later at the TRPA Governing Board meeting, agency staff called for "emergency action over New Zealand mud snails, bass, bluegill, Eurasian watermilfoil, and quagga mussels." A resolution read, "Zebra and quagga mussels and New Zealand mud snail pose a major threat to Lake Tahoe and other lakes of the Lake Tahoe region." Erroneously claiming that "the invasive mussels have wreaked billions of dollars' of havoc in Midwestern waterways," the Governing Board granted the emergency action. Regan said the resolution was largely symbolic, but by declaring a state of emergency, new doors for funding would open.

Formed a decade before quagga were found in Nevada, a cooperative effort by local, state, and federal agencies was organized by U.S Fish and Wildlife to stop the western advance of quagga and zebra mussels. Named the 100th Meridian Initiative, the agency devised a plan of preventing mussel dispersion by trailered boats. Educational materials, watercraft inspection and quarantine guidelines, and training programs were developed, and featured a gallery of photos showing clusters of mussels attached to pipes and boat propellers. In recent years, they were bolstered with staged photos of tennis shoes caked with mussels, and a photo of a mussel-covered shopping cart is ubiquitous.

Even though quagga were transported west of the Rockies by the Colorado River, in California trailered boats were given special attention at state-line agricultural inspection stations, where mussels became the primary species of concern. Claiming "the two mussels are invasive species that disrupt lakes' food chains, clog water pipes, and damage boats, docks, and ramps, and threaten Lake Tahoe, among other lakes in the state," the California Department of Food and Agriculture asked for an additional $2.5 million to inspect boats for mussels at border inspection stations. Without closing lakes, checking boats for mussels was the most proactive way agencies could demonstrate that they were serious about stopping the spread of quagga.

It is often perceived that aquatic animals only need water and food to survive. But as any kid who has rescued minnows from the bait bucket will attest, that isn't true—the right mix of aeration, temperature, nutrients, and cover is required to get the little fish through only a few days. In freshwater environments, habitat and climate are critical to survival. No one would suggest that Lake Tahoe could be infested by alligators from Florida, but hydrilla from Florida or mollusks from Eastern Europe aren't as intuitively understood. And if a trusted agency says a non-native plant or animal can invade a water body and be ecologically devastating, they are readily believed.

Early fears that zebra mussels would wreak havoc on Great Lakes' water-intake infrastructure weren't realized and effective and inexpensive control methods were developed for infested facilities. Even before the May 2007 conference in Incline Village, TRPA had launched a public information program that focused on the mussels and the damage they might do. Using materials from the failed 100th Meridian Initiative, TRPA adopted the program's message to "Clean, Drain, and Dry" boats before launching. The federal agency, which was organized in 1969 to oversee Tahoe Basin land use and to protect and preserve Tahoe's environment, conducted training sessions for volunteer inspectors, information on quagga was distributed, and random inspections began at Tahoe's public launch sites during the 2007 boating season.

Often called exotics or aliens, non-native species are introduced into aquatic habitats by such innocuous means as kids freeing aquarium fish or fishermen dumping bait boxes. Most introduced species soon die because they are biologically unfit for a foreign habitat or are eaten by predators. Some survive, reproduce, and carve out a niche in their new habitat, and a tiny few flourish. If a non-native species causes human health issues or measureable ecological or economic damage, it is invasive.

Ecologists are divided on the magnitude of environmental threats presented by non-native species. A *New York Times* article published a year after quagga were found in southern Nevada said, "Exotic species receive lots of attention and create lots of worry. Some scientists consider biological invasions among the top two or three forces driving species into extinction. But researchers argue that attitudes about exotic species are too simplistic. ... They [seldom] set off extinctions [and] they can spur the evolution of new diversity.

"'I hate the exotics-are-evil bit—it is so unscientific,' said Dov Sax, an ecologist at Brown University. And he argues 'that competition from exotic species shows little sign of causing extinctions.' This finding is at odds with traditional concepts of ecology. Ecosystems are often seen as having a certain number of niches that species can occupy, and once an ecosystem's niches are full, new species can take them over only if old species become extinct. 'But as real ecosystems take on exotic species, they do not show any sign of being saturated,' Sax said."

In a paper published in *The Proceedings of the National Academy of Sciences*, Sax analyzed the rise of exotic species on six islands and island chains. He concluded that non-native plants became naturalized at a steady pace over the last two centuries, showed no sign of slowing down, and, in fact, the total diversity of the islands had doubled.

Those who champion the "diversity of species is good" school of thought point to New Zealand, a country celebrated for its lush vegetation. Similar to America, when Europeans began arriving there two centuries ago, "they brought with them alien plants—crops, garden plants, and stowaway weeds. Today, 22,000 non-native plants grow in New Zealand. Most of them can survive only with the loving care of gardeners and farmers. But

2,069 have become naturalized[2] and spread across the islands on their own. Now there are more [non-native] plant species in New Zealand than native species.

"Fish also show this pattern," says James Brown of the University of New Mexico. He said that whenever he visits a river where exotic fish have been introduced, "I ask, have you seen any extinctions of the natives? The first response you get is, not yet, as if the extinction of the natives is an inevitable consequence. There's this article of faith that the net effect is negative."

Brown doesn't think that faith is warranted. For example, in Hawaii 40 new species of freshwater fish have established, and the five native species are still present. Brown and his colleagues argue that because they compete better than the invasive species in most habitats, native species aren't becoming extinct.

Brown says that huge negative effects of invasions are not documented in the fossil record, either. "You see over and over again that this is never the case," he says. "The overall pattern almost always is that there's some net increase in diversity. That is because these communities of species don't completely fill all the niches. The exotics can fit in there." (Zimmer, 2008)

Eight years later, the *New York Times* revisited the issue. Dov Sax said, "I think the dominant paradigm is still 'when in doubt, kill them.'" But a growing number of scientists are challenging this view, and say that most invasive species aren't destructive and many are beneficial. Ken Thompson, an ecologist and retired senior lecturer at the University of Sheffield in England says, "It's almost a religious kind of belief, that things were put where they are by God and that that's where they damn well ought to stay. [But] we're actually moving plants and animals around the world all of the time and have been for centuries."

---

[2] Non-native species that propagate without human intervention

Along with a growing number of invasion ecologists, Thompson notes that controlling the spread of non-native species is virtually impossible in an era of globalization. And as warming temperatures and resulting climate change pushes more species out of their home ranges into new areas, the number of so-called invaders is likely to multiply exponentially. (Goode, 2016)

Those who disagree with the "diversity is good" school of thought point to environmental and economic harms done by aquatic species such as the sea lamprey, dreissenid mussels, and Asian carp. Silver, grass, black, and bighead carp—the fish called Asian carp because of their continent of origin—are an example of how some aquatic animals find niches that natives are unable to occupy. The closely related filter-feeding carp were imported by catfish farmers in the 1970's to reduce algae in aquaculture ponds and have since migrated throughout the Mississippi River system, feeding on the rivers' plentiful algae and plankton, the foundation of the aquatic food chain. (Watson, 2010)

Accusations that Asian carp are degrading water quality in the Mississippi and causing extinctions are wrong. The hardy fish are able to survive where others can't. Before the Army Corps of Engineers built 29 dams along its 1,250-mile length during the Depression to facilitate commercial navigation, the Upper Mississippi ran as clear as a mountain stream and supported more than 200 native fish and mollusk species, and its wetlands housed more than 300 varieties of birds. The dams destroyed the wetlands and transformed the river into a string of muddy reservoirs, halting natural flows that for eons had produced a dynamic environment and vibrant wildlife habitats. (Walker, 2010)

Today, more than 140 fish species are still found in the Upper Mississippi, but only about half are native, and below the Wisconsin River, sport fishing is almost non-existent. The *Upper*

Mississippi River Fisheries Plan 2010 says the first non-native fish in the Upper Mississippi "pre-dates 1883 when common carp were introduced. ... Like many exotics, once introduced into a new environment, [carp] became detrimental to native populations." That too is wrong. It wasn't carp that wiped out the trout that they shared the river with for more than a half-century—it was the Army engineers who destroyed the trout's habitat.

Sorting out undesirable non-native plants and animals is difficult. What's invasive in one country or region may be prized in another. In California millions are spent on improving the Sacramento-San Joaquin Delta habitat and flyway for Canada geese, but in Europe the North American goose is considered a biofouler and is listed as one of Europe's Top 10 invasive species. In Asia golden clams are thought to bring prosperity, are used for medicinal purposes, and are sold to keep aquariums clean. In North America they are called Asian clams, and at Tahoe millions have been spent to eradicate them. In the South, bullfrogs are farmed for their tasty leg meat and are served in fine-dining restaurants. At Tahoe the U.S. Forest Service wants to eradicate them using the Canadian-made Electro-frogger and by harvesting native plants to destroy their habitats and prevent repopulation.

Categorizing non-native aquatic species is equally difficult, and although the words have significantly different definitions, "invasive" is often used when "nuisance" is preferable or when "naturalized" is correct. The federal Aquatic Nuisance Species Task Force defines aquatic "nuisance species" as "nonindigenous species that threaten the diversity or abundance of native species or the ecological stability of infested waters, or commercial, agricultural, aquacultural or recreational activities dependent on such waters." The task force defines an invasive animal or plant as one that is "non-native to the ecosystem, and whose introduction causes economic or environmental harm or is harmful to human health." At the 2015 Annual Meeting of the Western Regional Panel on ANS, University of Texas-Arlington professor, Bob McMahon, conducted the conference's liveliest session: "When is a species invasive?" Eighty of the nation's top AIS managers couldn't forge a better definition than the task force's.

Because few species cause economic damage and because environmental harm is difficult to quantify, to advance their own agendas some academics and researchers redefine "AIS." In his paper, *Surveillance and Control of Aquatic Invasive Species in the Great Lakes,* Jake Vander Zandem of the University of Wisconsin defines AIS as: "Any aquatic species that has been transported beyond its native range, regardless of adverse impact." (Vander Zandem, 2007) His broad definition is usually reserved for "exotics," and by his definition, brown trout, imported from England in the 19th century; mackinaw from the Great Lakes; and even kokanee salmon and rainbow trout that are native to northern California would be an invasive species at Lake Tahoe.

Rainbow trout are an example of how blurry distinctions are between invasive, nuisance, and naturalized species. Dubbed by some as a "synthetic fish," wildlife agencies annually release about 2 billion rainbow trout for sport fishing, and agencies have stocked fisheries in every state and 49 countries with the feisty fish. Using the same explanation that some ecologists use to justify controlling non-native species, others point out that rainbow negatively impact native fish by out-competing them, eating their eggs, and preying on their fingerling.

At Tahoe, where no species introductions have caused any measureable economic or environmental harm, AIS program managers scramble to justify classifying bluegills, bullfrogs, Asian clams, and bass as invasive species. To open the 2015 Tahoe AIS

Forum, Nicole Cartwright, AIS Program Coordinator at the Tahoe Resource Conservation District (Tahoe RCD) asked the audience of 40 agency managers and interested citizens what the definition of an aquatic invasive species is. When no one volunteered an answer, Cartwright rejected the official definition and invoked Vander Zandem's and stated, "An invasive species is any non-native species. The three you hear most about [at Tahoe] are New Zealand mud snails and zebra and quagga mussels"—none of which are in the lake or could survive there if they were.

TRPA justifies the intentional introduction of non-native game fish by saying, "The nature of the relationship between non-native species and the local landscape is largely based on potential harmful impacts versus societal benefits. That is, society may deem [that] the benefits of *intentional* introductions of non-native species outweigh potential or realized harmful impacts."

TRPA has scrapped standard definitions and unilaterally decides which nonnatives are beneficial or labeled as AIS, and its decisions are inconsistent. Biologists at a North Tahoe hatchery accidentally introduced kokanee salmon into the lake in 1944. They are now the centerpiece of the Forest Service's Taylor Creek Stream Profile on Tahoe's southwest shore and are celebrated every October during the Fall Fish Festival when crimson-bellied kokanee put on a memorable show as they race up Taylor Creek to spawn. But naturalists say the festival showcases "artificial nature" at the expense of the natural environment.

In addition to bisecting 400 acres of Tahoe's second largest wetlands with roads and trails, to prevent flooding and so that the kokanee can return to where they hatched, the Forest Service removes beaver dams along Taylor Creek. For years the removal was justified because beavers were believed to be non-native to the Basin. However, radiocarbon dating of buried beaver-dam wood verifies that beaver inhabited the eastern slope of the Sierra until they were trapped to extinction in the 19th century.

A Taylor Creek habitat study convincingly postulates that beaver dam removal decreases the creek's wetlands and their effectiveness and increases stream flow and algae-producing phosphorus—all factors that negatively impact Tahoe's clarity and water quality. (Muskopf, 2007) Removal of the beaver dams also thwarts efforts to reestablish Lahontan cutthroat trout. Deep pools that were created by the dams used to provide critical winter refuge for the trout, and unlike the smaller kokanee, cutthroat trout can easily move across beaver dams. In early 2015, the Sierra Wildlife Coalition began painting tree bases in sensitive areas with a sand-latex paint to discourage the beavers from gnawing at them. Initial results were positive.

The Coalition's sensitive approach to beaver repopulation at Lake Tahoe is more the exception than the rule. More common is the situation at Scotland's River Tay. After being hunted to extinction in Britain centuries ago, beavers returned to the River Tay. Some residents viewed the beavers as destructive interlopers and insisted that because the beavers were absent for centuries, their native status was nullified.

"It's just silly," said Sheffield's Ken Thompson of the reaction to the Scottish beavers. "I don't think we would have ended up in this ridiculous situation if we weren't so bombarded by propaganda about invasive species."

Mark Davis, biology chairman at Macalester College, says, "Assuming that exotic species are inherently bad, that ecosystem 'integrity' can be measured by the number of alien species, or even that newly arrived species are functionally different from longtime residents, simply isn't supported by science. Many introduced species are valuable and useful." (Breining, 2009)

Davis writes: "Some nonnatives cause problems, some produce desirable effects, and most we don't think much about. When it comes to [species] management, it is important that we worry about the right things. We should always worry about species that threaten health, affect the economy, or undermine ecological services. We should worry much less about species that aren't producing any of these effects but simply alter [a habitat's] composition. [Environmental managers] simply do not have the luxury to consider as harm mere ecological *change*."

In 2011, Davis and 18 other researchers submitted an article to the journal *Nature* entitled, *Don't judge species on their origins*. In the peer-reviewed article they argue that non-native organisms should be assessed on their environmental impact rather than on whether they are native. The researchers wrote: "Over the past few decades, 'non-native' species have been vilified for driving 'native' species to extinction and polluting 'natural' environments.

"Intentionally or not, such characterizations have created a pervasive bias against alien species that has been embraced by conservationists, land managers, policy-makers, and the public, as well as by many scientists. Increasingly, the practical value of the native-versus-alien species dichotomy in conservation is declining, and even becoming counterproductive. Yet many still consider the distinction a core guiding principle." Predictably, the reaction from land managers, conservationists, and invasion biologists was that Davis and his colleagues were wrong.

"All I argue for is a more nuanced characterization of what's been happening," says Davis. And he says that he's finding traction for his ideas: "People are thinking more carefully about the words they are using, the assumptions they might be bringing in." In his textbook, *Invasion Biology*, Davis writes: "If [nonnatives] are only changing the ecosystem but not causing significant harm, then altering one's perspective is certainly much less costly than any management program.

"Overselling the threat [nonnatives] pose is bound to lose credibility as exotics make up ever more of the biota around us," says Davis. "And it will lead to misguided spending on projects ... when we should focus instead on disease organisms and other more pressing threats. It's very important to distinguish harm from change." (Breining, 2009)

A 2013 *Acta Oecologica* paper on invasive species points out how many species that are labelled as "invasive" are, in fact, beneficial and says: "A number of [nonnatives] are actually positive for ecosystems and human well-being. For instance, [zebra and quagga mussels are] a biofouler of pipes and boats, but they can improve water quality through their filtration of nuisance algae, a valuable effect that is often overlooked. Many of the most damning claims about invaders are not backed up with hard evidence. This might be skewing priorities when it comes to dealing with them. Some invasive species are getting a harder time than they deserve. It's an emotive subject but it needs to be looked at in a balanced way. ... We could find little evidence of the reported [negative] effects. If this is the literature behind the worst species, then what is the evidence like for others?

"We found that negative effects, particularly economic ones, were often assumed rather than quantitatively evidenced. ... In general, the evidence for impacts of these 'worst' invaders was severely lacking. We conclude that invasive species management requires prioritization, which should be based on informed and quantified assessment of the potential ecological and economic costs of species (both positive and negative), considered in the proper context of the invader and ecosystem." (McLaughlan, 2013)

At Tahoe, millions have been spent to control Asian clams, a species that does no harm there and is in fact ecologically beneficial—a 2008 AIS fact sheet produced by UNR that is still posted on TRPA's website states no harms done by Asian clams but lists their ecological benefits and how they improve water quality: the clams "filter particles suspended in water, including bacteria, algae, and detritus." And millions more are spent to protect the lake from a quagga mussel invasion that can't happen. Asian clam control and protecting the lake from invasive species by boat inspections are the centerpieces of Tahoe's AIS management plan, but there has never been a quantified assessment by experts to determine if any nonnative that Tahoe's boat inspectors look for can harm the lake's aquatic ecology or even survive there.

— 2 —

It Takes a Tough Species to Live at Tahoe

"We are so spoiled. We have no ticks, fleas, poison oak, poisonous snakes, thorns, and quagga and zebra mussels—yet." ~ Kristi Boosman, TRPA Partnerships Officer

Tahoe is often described as having "sensitive" or "fragile" ecosystems, which makes it vulnerable to invasion by non-native species. But the opposite is true: only the hardiest plants and animals can survive the Sierra's climatic extremes and scant food sources, and as outdoorsmen know, the reason the mountains have few species is that the rugged habitats are inhospitable to most. When the last glaciers melted 13,000 years ago, other than microorganisms, the 400-mile-long Sierra Nevada range was scoured of plant and animal life. Slowly, cycles of mosses, lichens, sedges, cushion plants, and grasses filled in mountain ledges and spread across alluvial valleys creating soil for hardy conifers and forming spectacular meadows. And in Lake Tahoe seven fish species evolved and slipped into ecosystem niches.

The glaciers left behind jagged peaks and more than 2,000 crystal-clear lakes and ponds that still remain virtually plant-free.

Frigid, granite-slab and gravel-bottomed lakes, drenched with searing sunlight, provide a harsh habitat for aquatic animals, and the extraordinarily thin web of fish that evolved in Tahoe developed biological traits that allowed them to prosper in cold, low-nutrient water. Three species of minnows, three medium-sized fish, and at the top of the food chain, the Lahontan cutthroat trout (earlier known as silver trout), which migrated from the Great Basin's Lake Lahontan thousands of years ago, filled the ecological slots reserved for fish.

John Fremont, the explorer who discovered the stunning lake in 1844 called the giant strain of trout, which grow to more than 40 pounds, "salmon trout," and said of them, "Their flavor was excellent—superior, in fact to that of any fish I have ever known." And when he and Kit Carson ascended the Truckee River to Lake Tahoe, they named it the Salmon Trout River. The exceptionally slow-growing, long-lived fish fed on minnows and insects, and for millennia cutthroat trout were the undisputed kingfish of the Sierra Nevada.

During the mid-19th century, Comstock miners, railroaders, and Tahoe loggers dined on the tender fish and depleted their numbers, but it was the 1887 introduction of mackinaw, a freshwater char called "lake trout," into Tahoe that drove the lake's cutthroat trout toward extinction. The fast-growing char gobbled up immature cutthroats and introduced a parasite that ultimately shoved Lahontan cutthroats from their niche at the top of Lake Tahoe's food chain. Ironically, mackinaw didn't begin to prosper until the 1960's when large-scale development on the lake's south shore destroyed wetlands, allowing nutrient-enriched water from the Upper Truckee River, creeks, and storm runoff to flow freely into the lake.

"Oligotrophic" is the term scientists use to describe water that is pure and low in nutrients and plant life but still rich in dissolved oxygen, one of the common essential elements for all animal life. Because of its exceptional water quality, clarity, and purity, ecologists rank Lake Tahoe as an "ultra-oligotrophic" lake, and it ranks behind only Oregon's Crater Lake as America's clearest lake. In an environmental context, "water quality" refers to the physical, biological, and chemical characteristics of water.

Environmental awareness has come a long way since the 1860's when Tahoe forests were clear-cut to provide timber for San Francisco Victorians, railroad ties, and timbers for Comstock mines. Then, sediment in runoff from storms and spring snowmelt turned Tahoe's nearshore a murky brown, but until almost a century later the lake's primary pollutant was human waste—effluent was allowed to drain and seep directly into the lake. In 1951 septic tank business owner Lester Nagy put himself out of business by forming a modern collection and treatment system along South Tahoe's Highway 50 corridor.

Seven years later a University of Michigan Ph.D. on his way to the University of California, Davis (UC Davis) to assume his first professorship saw Lake Tahoe for the first time. He had no idea that he would become the catalyst for science that would secure the lake's environmental future. Less than a decade after arriving in California, Charles Goldman contributed the single greatest insight into how to preserve Lake Tahoe's exceptional water quality and clarity—he demonstrated that even chlorinated wastewater was detrimental to the lake's health. By showing skeptical civil engineers that adding treated effluent to Tahoe water quickly turned the crystal-clear water green with algae, Goldman convinced community and government leaders that wastewater should be pumped out of the Tahoe Basin.

For the next half-century, Goldman taught a generation of Tahoe scientists and aquatic researchers the intricacies of the lake's ecology and directed its agencies to adopt solid, science-based policies and meaningful environmental protections. In 2001 he capped an exemplary academic and scientific career when he convinced President Bill Clinton and Vice President Al Gore to sign onto a plan that annually channeled $100 million from federal and state government into the Tahoe Environmental Improvement Program. After attending a private seminar aboard Goldman's research boat on the state of the lake's health, the President of the United States described the outing and joked about his vice president's fondness for environmental science: "When I saw the vice president this morning, he was virtually glowing, and I knew he had been here in his element," Clinton said. "And the minute we got on the boat, I got my Marine Biology 101 lecture—about phosphorus, nitrogen, and what does what. I mean, I could pass anybody's test now."

Aquatic species, such as Tahoe's native fish and mollusks that can survive in low-nutrient conditions, are called oligotrophs. They are cold-blooded animals that are characterized by low rates of metabolism and growth that allow them to compete favorably for the sparse food found in clear, cold water. Free-floating plankton form the bottom layer of the food chain and are the primary food source of most oligotrophs. Because oligotrophic water has scarce plankton, the base of Tahoe's aquatic food chain is anemic and doesn't support most of the "warm-water fish" and mollusks that are found in North America's temperate zones. Nonnatives that survive in Sierra Nevada lakes seldom cluster in dense colonies, and often they are stunted in size.

Of the hundreds of North American freshwater mollusks only a few are physiologically adapted to live in oligotrophic lakes. By siphoning large quantities of water, the hardy few filter enough microscopic plants and animals to sustain their remarkably efficient bodies. One of the most successful oligotrophs is the river mussel that is found in the Upper Truckee River. It has a lifespan that can reach 50 years. Two native mussels, the California floater and the western pearlshell, inhabit Tahoe's clear waters, along with the ramshorn snail and the montane peaclam.

The small mollusks were among Tahoe's first residents, and sometime in the last half-century they were joined by a foreign cousin, the Asian clam. Immigrants, who brought the small clams with them to California during the Great Depression, called them golden clams, because they were believed to bring good luck. In addition to being harvested for food, they were used for game fish bait and to keep aquariums clean.

Before embarking on a control strategy for any aquatic invasive, it is important to examine habitat characteristics and evaluate potential harm to the habitat. If the economic and/or environmental cost to protect environmental or infrastructure assets is greater than the damages, controlling the invader is unwarranted. Typically, newly introduced species achieve high population levels until seasonal changes or climate or food cycles cause stabilization at lower levels. This was the case when Asian clams invaded the Colorado River over 40 years ago. Today, they are sparse, are harmless, and are virtually unnoticed. (Strayer, 1999) A near identical situation occurred at Donner Lake—and likewise is developing at Lake Tahoe.

Because of exceptionally low calcium and scant plankton and algae, Asian clams in Sierra lakes are stunted and only grow to about a quarter of the size of those in warmer, high-nutrient water. At Donner Lake, the tiny clams propagate, but don't proliferate, in water where the calcium concentration is half that required by quagga. Adults tolerate temperatures as low as 36°F but don't reproduce below 62°F. Besides restricting growth, water temperature also affects Asian clams physiologically, including a longer time to mature and a shorter lifespan. (Rosa, 2012)

In the warm, sandy-bottomed shallows along Lake Tahoe's South Shore, Asian clams find a hospitable habitat and populate in densities up to 6,000 per square meter. That seems like a lot, but in an ideal habitat, such as the San Francisco Bay, Asian clams grow up to six times the size of Lake Tahoe's clams and form colonies ten times larger. There conditions are near-perfect for them, and they have changed the bay's benthic[3] ecosystem, and impacted some fish, especially delta smelt (Martin, 2013). Asian clams in San Francisco Bay underscore the dilemma invasives pose in large bodies of water. The 1,500 square mile estuary has been undeniably affected by the clams, but they are only one of 250 non-natives to populate the bay, and they can't be controlled.

When projecting harms caused by Asian clams, damage in optimal habitats like San Francisco Bay is often transposed onto marginal or poor habitats like Tahoe's, where they have caused no harm or even shown a potential to harm. The bay's habitats are dramatically dissimilar to Lake Tahoe's ecosystems. San Francisco Bay is North America's largest and most bio-diverse estuary, and it supports more than 500 major aquatic plant and animal species— Lake Tahoe supports fewer than 40 aquatic species, and similar to the Bay, half of them were introduced.

Although at nearby Donner Lake, Asian clams were sold for bait until the 1980's and have been found in the lake's warm, gravelly shallows since then, they weren't noticed at Lake Tahoe until 2002. Now believed to have been in the lake much longer, the tiny clams weren't commented upon until UC Davis and UNR researchers performed a study, *Asian clam of Lake Tahoe: Preliminary scientific findings in support of a management plan.*

---

[3] The bottom or sediment under a body of water.

The study cited no costs or harms caused by Tahoe's clams, but it concluded: "The Asian clam is an aquatic invader. Because of its fecundity and ability to thrive under a range of conditions, it is considered one of the worst freshwater invasive species."

Triggered by the discovery of quagga in southern Nevada reservoirs, Tahoe academics began characterizing all nonnatives as invasives, and their report proclaimed: "The rapid expansion of Asian clams in one year combined with their demonstrated potential to alter the ecology of the lake via unprecedented levels of algae in the near-shore represents a major new threat to Lake Tahoe." As later analysis by the same researchers confirmed, the statement was wrong: by consuming algae, clams decrease it—a fact known and promoted by TERC and UNR researchers as early as 2008. And the clams didn't rapidly expand in one year—before then, no one was looking for them.

Because the clams had caused no harm, the plan received little attention. That changed a year later when the academics claimed the clams might be a "facilitator" for quagga mussels. The *Reno Gazette-Journal* reported that limnologist, Sudeep Chandra, the director of UNR's Aquatic Ecosystems Laboratory, said: "It might be this existing invader, [Asian clams, are] modifying the bottom environment. One invader can facilitate another." The news article added that "officials at Lake Tahoe are increasingly concerned that quagga or zebra mussels, already thriving in parts of California and southern Nevada, could become established in the lake. The quickly reproducing mollusks could profoundly alter the lake's ecosystem, clog water intakes and litter now-pristine beaches with sharp and stinking shells. 'I think it's a pretty serious issue,' said Mara Bresnick, chair of the TRPA Governing Board. 'It could wreak havoc.'

"Chandra recently wrapped up the first phase of a risk assessment into the potential mussel invasion of Tahoe. The work involved taking water samples from 83 locations along Tahoe's shoreline and measuring the water for calcium. The next stage of research will involve laboratory tests to determine if mussels can survive in water taken from Tahoe. Chandra said, 'Preliminary results of the assessment's first phase are somewhat encouraging. There is some variability but overall, calcium levels seem relatively low. But the few locations where calcium levels were higher are also in some of the same areas where Asian clams are found. That raises the possibility that as clams grow and die, they leach out calcium into the water. That could pose a danger that quagga or zebra mussels, if introduced in those areas, might find sufficient calcium to gain a toe-hold in the lake. It creates a little hot bed zone. Elevated calcium levels could be favorable to the quagga.'"

In August 2007, Sudeep Chandra was an undistinguished UNR associate professor of limnology. He had received his B.S. and Ph.D. degrees from UC Davis and had participated in limnology studies at Lake Tahoe for almost 20 years, but he had performed no research on aquatic invasive species or participated in any AIS studies. In less than a year, he became the regional authority for the nation's most aggressive AIS programs.

Chandra's theory that clam shells elevate calcium levels in microzones by leaching it into the water column is astoundingly incorrect, and is so patently wrong that it is remarkable than none of his colleagues bothered to correct it. And it is still frequently cited today by Tahoe agencies as a justification for controlling Asian clams. The bogus theory was the first of many incorrect premises that Chandra used to vilify non-native aquatic species.

Mollusks filter calcium to build their shells and for energy and muscle function. As they grow, secretions of proteins and crystalline calcium carbonate expand their shells in layers. Bivalve mollusks (those with hinged shells) filter available calcium ions from the water. Called the edaphic source, dissolved calcium in lakes is primarily from weathered rocks and runoff. Secondary sources of calcium are from atmospheric dust, pollution, and concrete leaching.

It would seem that by filtering calcium and concentrating it into shell material that clams would create stockpiles of calcium that would concentrate the element in an area. But most mollusk shells are more rigid than concrete and take longer to decompose. For dissolution, calcium carbonate needs slightly acidic water and minerals to combine with. Consequently, the purer the water, the longer shells take to decompose. In Lake Tahoe, mollusk shells take decades to disintegrate. (McMahon & Bogan, 1991) And the reason Asian clams are found in areas of Sierra lakes where algae and calcium are highest is because, like all animals, they prosper where food and nutrients are most plentiful.

Californians know there is enough water in Lake Tahoe to cover their state to a depth of 14 inches. That's a lot of water—36 cubic miles of water. High school chemistry courses teach how to calculate the amount of calcium needed to raise Lake Tahoe's dissolved calcium from its current average of 9.16 parts per million (ppm)[4] to 12.0 ppm—the minimum level needed for quagga propagation. Students easily determine that it takes about 110,000 tons of calcium to raise the lake's calcium concentration by 2.84 ppm. That's enough pure calcium to fill more than 1,150 railcars. They also know that clams don't increase total water calcium, and by concentrating it into shell material, they slightly decrease water calcium concentrations—and that the total volume of Tahoe's calcium remains near constant.

Similar to mussels, clams are filter feeders, and along with algae and nutrients, they measurably improve clarity and water quality by steadily removing pollutants and bacteria. A 2012

---

[4] Parts per million (ppm) are equivalent to milligrams per liter (mg/L)

research study, *Long-term trends of Great Lakes major ion chemistry*, reports that U.S. EPA and Canadian scientists attribute slightly decreased calcium ions in the Great Lakes to zebra and quagga mussels. (Chapra, 2012) It is wrong to say that Asian clams facilitate quagga mussels, but at Tahoe this erroneous correlation is accepted as fact. Worse, the unfounded fear that Asian clams could increase the lake's calcium to levels where quagga mussels could establish became the foundation for an incredibly expensive and environmentally destructive control program.

Worldwide, there are more than 10,000 clam and mussel species, and mussels are an aquatic animal family found in every large water body. They have been on earth for 250 million years and prosper in every aquatic habitat that doesn't freeze. Certain species of marine mussels prefer quiet bays, others attach to wave-washed rocks and thrive in pounding surf, and others flourish at extraordinary depths near hydrothermal vents. A broad range of more than 300 freshwater mussel species are found in every large water body in North America, and besides improving water quality, they are a critical link in the freshwater food chain.

An example of mollusks ecological importance is illustrated by the eastern oyster. Oysters are saltwater clams, and they were once so plentiful in Chesapeake Bay that they filtered a volume of water equal to the bay's every three days. They are now only one percent of historic populations, and the bay's water quality is at an all-time low.

Evolutionary biologists know that as animals evolve they maximize available food in distinct habitats and that only a slight change in environmental conditions can cause migrations or even extinctions. Because non-native species didn't evolve with the natives' ability to acclimate to seasonal changes in the food web or to cyclic climate and water quality changes, aquatic aliens usually can't favorably compete with natives over the long term—in much of North America, dreissenid mussels are an exception.

Zebra and quagga mussels are native to Northeastern Europe and evolved to survive in a broad range of freshwater conditions. After they were discovered in the Great Lakes in the late 1980's, in order to learn how to prevent damage to water plant infrastructure, aquatic biologists raced to examine the mussels biological tolerances, environmental thresholds, water quality requirements, and for methods to control them. Extensive research uncovered a range of mechanisms. Chlorination is the most common treatment. It was found that small amounts of chlorine introduced at the mouth of water intake pipes kill dreissenid larvae (called veligers) and effectively protect pipes and internal infrastructure.

After introduction, the potential of a species to survive in an environment depends on its biological requirements and their availability in a habitat. These life-sustaining elements and habitat conditions are called "chalk variables." To determine the invasion threat posed by potential nonnatives, through experimentation and observation of similar habitats research ecologists measure minimum and maximum chalk variables to determine survivability thresholds of essential elements and environmental factors. Affected by alkalinity (pH) and water temperature, the chalk variable most likely to determine the survival of zebra and quagga mussels is dissolved calcium. (Mackie & Claudi, 2009)

A quarter-century of extensive analysis of environmental data of dreissenid infestations and laboratory testing established the levels of calcium concentration at which mussels survived, colonized, and thrived. It was conservatively established that adult quagga need average dissolved calcium of at least 10 ppm to survive more than two months, and their veligers require concentrations greater than 12 ppm to survive more than a few days. Predicting the ability of a nonnative to merely survive is much easier than predicting its ability to infest a habitat, but several studies correlating colony density to calcium levels established that quagga never colonize in harmful numbers in open freshwater until calcium exceeds 20 ppm.

Dissolved calcium in Sierra lakes is less than 75 percent of that which dreissenidae need for long-term survival and 50 percent of that needed for harmful colonization. When quagga were found in Lake Mead, further analysis of background levels[5] of water quality wasn't necessary—muscologists knew that the Sierra wouldn't be invaded by dreissenids. But instead of rejoicing over the region's natural good fortune—and as if they ignored the rich body of science—TRPA began a relentless campaign to justify their own hype over AIS threats to Lake Tahoe. Only months after quagga crossed the 100th Meridian, the agency mobilized the environmental community, volunteers began inspecting boats for aquatic invaders, and TRPA developed an AIS management plan.

Chandra's erroneous claim that Asian clam shells could increase the dissolved calcium in parts of the nearshore was seized by TRPA as a plausible way to explain how an immense waterbody that was well below the calcium survival threshold of quagga could support them, and TRPA's public information staff proclaimed that Asian clams "facilitated" quagga. Tahoe media was peppered with press releases and opinion pieces assailing the clams, and in addition to claiming that they created "calcium hot zones," TRPA's supporting research group of UNR and UC Davis academics, said Asian clams were responsible for causing algal blooms and decreasing water quality and clarity—claims they quietly retracted three years later.

---

[5] The concentration of a substance that occurs naturally

In the spring of 2008, few in the Tahoe Basin had heard of Asian clams or quagga mussels, and those who had were often confused between the two and were unclear as to how they could cause catastrophic damage to the lake. Prior to the 2008 boating season, the *Tahoe Daily Tribune* ran a four-part series on invasive mussels that began: "Half a billion dollars spent on water quality and environmental improvement programs. A century of debate about how to best protect the blue waters; a bi-state compact and three regional plans to protect the fragile ecosystem; and one mussel, brought in through the crevice of a boat, could destroy every effort over the past 50 years to keep Lake Tahoe blue.

"TRPA Executive Director John Singlaub said, 'They will jam up water intakes, shed their shells, and the beaches won't be useable, they put drag on boats so boats have to constantly be cleaned. It could kill the economy and ruin recreational tourism based on the lake. This should be of concern to everyone.'

"Quagga and zebra mussels are close cousins, members of the dreissena family of shell fish. While the differences between the two are subtle—from appearance to ecological tolerance—the effect is the same. Once introduced, these filter-feeders will eat the food and nutrients species higher on the food chain need to survive. In turn, they can destroy an ecological system and collapse an entire food web. 'If you remove the basis of that food web everything dies,' said David Britton, the ANS coordinator for the southwest region of the Fish and Wildlife Service.

"Jason Roberts, an environmental scientist with California Fish and Wildlife, said, 'Ironically, the introduction of quagga or zebra mussels to Lake Tahoe could very well increase lake clarity, But that clarity would come at the price of a destroyed ecosystem and a bloom of blue-green algae—which could turn the water different colors and cause it to take on a bad smell and taste. It's not a thing you would want in the lake.'

"The locust-like creatures spawn continuously if conditions are amenable—some adult mussels can produce a million eggs in one spawning season. Once their life cycle is over, the mussels die off in large batches. Their sharp shells can saturate beaches. 'You have accumulated mussels dead in the water all at once that creates quality issues, and have been known to have recreational impacts,' said Ted Thayer, the natural resource and science team leader for TRPA. 'They can attach to most surfaces with their byssal threads—including boat hulls, water intakes, lake floor, and even other native mussels.'

"In their larvae-like stage, quagga and zebra can find safe haven in standing water in the bilges, live-wells, and motors of boats and be easily transported from one area to another. If they grow into water-treatment facilities, quagga and zebra mussels can clog pipes and cause millions of dollars in infrastructure costs.

"U.S. Congressional researchers estimated that the Great Lakes zebra mussel infestation cost the power industry $3.1 billion in the 1993-1999 period, with an economic impact to industries, businesses, and communities of more than $5 billion. 'In North America they have no native predators or natural controls, so a population can grow rapidly,' Thayer said.

"'They are the poster-child for invasive species,' Britton said. And once introduced to a body of water—one mussel larva, smaller than a grain of sand—can begin an infestation that cannot be eradicated, only contained. 'The best we could hope for is control,' Thayer said. 'Once you have them, you've got them.'" (Flanzraich, 2008)

While overlooking the most critical ecological fact—that quagga can't survive in Tahoe water—every environmental threat or economic claim stated in the article was inaccurate, irrelevant, or grossly exaggerated. The species that TRPA still says are "a major threat to Lake Tahoe and other lakes of the Lake Tahoe region" were never found in the Tahoe region or anywhere else in the entire Sierra Nevada range.

The fear of quagga invading the lake so distorted public perception that all objectivity about nonnatives was cast aside. An indication of how ingrained TRPA's misinformation is on Tahoe public consciousness is that if the 2008 *Tahoe Daily Tribune* series were published today, there would be almost no protest about its inaccuracies, and it remains the most extensive reporting on quagga at Lake Tahoe.

An August 2008 *Reno Gazette-Journal* article reported: "Scientists from UC Davis Tahoe Environmental Research Center (TERC) are studying the possibility there may be a connection between Asian clams and this summer's algal bloom in Marla Bay, located just south of Zephyr Cove in an area where the clams are located. 'Tests have yet to determine any link, said UC Davis researcher Brant Allen.'" Because they feed on algae, detritus, and other microscopic plants and animals, no link could ever be determined to exist between Asian clams and algal blooms. Asian clams are sold to keep aquariums clean and the water clear.

Bivalves remove a broad range of biomass from the water column, and although clams and mussels' fecal matter contains nitrogen and phosphorus, elements that fuel algal growth, the volume of algae-promoting nutrients filtered from the water is significantly greater than the amount added. The major portion of the filtered biomaterial serves no useful function to the organism, but serves a beneficial function to the ecosystem (Ostroumov, 2005). And some scientists believe that dreissenidae are useful in certain environments to improve water quality. Cautioning that benefits and costs should be balanced on a site-by-site basis, *Cultivation of zebra mussels within their invaded range to improve water quality in reservoirs* concludes that zebra mussel cultivation offers an attractive tool for managing algae-infested and nutrient-enriched reservoirs (McLaughlan, 2013).

The same month that Allen told the Reno newspaper that Asian clams may have caused the Marla Bay algal bloom, the *Tahoe Daily Tribune* reported that researchers were trying "to determine if an unknown source of nutrients [was] sending nitrogen and/or phosphorus into the bay, fueling the [bloom]." TRPA spokesman, Dennis Oliver, said, "High fecal counts in the water near Marla Bay ... could be a potential source of food for the algae," and that the fecal matter may be from runoff with horse manure in it or from a sewer leak.

The most logical explanation for the Marla Bay bloom was given six months later when a similar bloom appeared on the North Shore in an area that is clam-free. Scott Hackley, a TERC researcher explained, "The lake is warming and goes through a period of mixing when it brings up nutrients." Hackley said that algal blooms can be fueled by sunlight, snowmelt loaded with nutrients, warmer temperatures, and many other factors. All evidence indicates the Marla Bay bloom was probably caused by nutrients in runoff, a sewer leak, or any one of a number of natural causes, but it certainly wasn't caused by Asian clams.

After boat inspections were stepped up from voluntary to "random, mandatory" for the 2008 boating season, the *Tahoe Daily Tribune* reported that TRPA spokesman Dennis Oliver said: "Mussels need approximately 12 parts per million of calcium to survive, [and] the lake's average calcium levels are about 10 parts per million." But he then incomprehensibly hedged the known science and said, "It is unknown if quagga mussels can thrive in Tahoe's calcium-poor environment."

As the 2008 boating season heightened, reports on the quagga threat to Lake Tahoe escalated from the risk of infestation being "encouragingly low" to Oliver saying, "The clam beds are a big concern. [Asian clams] will help introduce quagga mussels into the lake." And TRPA Executive Director John Singlaub, reinforcing that by saying the clams are a "larger problem than we thought." After the 2008 boating season, the TRPA governing board voted to require mandatory boat inspections to prevent quagga from infesting the lake. Then the AIS media campaign was ramped-up.

TERC Director Geoffrey Schladow spoke at the 2008 California Colloquium on Water and projected "What Tahoe Will Look Like in 2040." He said that, "In one year's time Asian clams have changed from being virtually unnoticed to having an all-consuming presence [and] have transformed [Tahoe's] pristine shore to an algal infested shoreline." In six months' time, Asian clams went from being totally inconspicuous to being blamed for infesting the lake's shoreline with algae and facilitating a potential quagga invasion. Seven years after Schladow told of the clams' all-consuming presence, most Tahoe beach goers still hadn't seen an Asian clam shell.

The same month Schladow told his UC Berkeley audience that Asian clams were transforming Lake Tahoe by nurturing algal blooms, TERC researcher Marion Wittmann told an Incline Village audience, "While the overall calcium level of Lake Tahoe is not substantial enough to support quagga or zebra mussel growth, areas near Asian clam beds may produce enough nutrients to support them. Quagga mussels need about 12 parts per million of calcium to survive and breed in a body of water. According to UC Davis and UNR research, the average near-shore calcium level of samples taken from 83 locations along the shoreline is 9.16 ppm. Tahoe is a low risk area for quagga. However, there might be localized areas that could facilitate the invasion of these species."

Schladow and Wittmann's presentations were made two months after TRPA passed mandatory boat inspections. Typically without confirming research, scientists don't speculate about what might happen or recommend multimillion-dollar programs to prevent a theoretical ecological event, especially when their speculations run counter to prevailing science.

Wittmann was almost correct when she said Tahoe is at low risk for quagga mussel infestation—it's at no risk—but her leap to saying there "might be areas" where they could survive was without scientific foundation. Without compelling data, casting doubt on accepted science exhibited either a lack of understanding of dreissenid physiology, water chemistry, and the environmental effect of the constant mixing of Tahoe's surface water due to wind and currents, or was extremely careless. In retrospect, TRPA and their aquatic science advisers' explanations to the media during 2008 appear to have been driven by the need to justify a hastily implemented boat inspection program.

In early 2009, the attack on Asian clams was stepped up. TRPA spokesman Dennis Oliver stretched baseless speculation into fact and said: "Researchers have found a connection between clam beds and an algal bloom in Marla Bay last summer. Increased calcium levels created by the clams could provide a foothold for other invasive species. That could open the door for the quagga and zebra mussels."

UPI Science News reported: "'Lake Tahoe is a place that people have tried hard to protect,' said John Reuter, an aquatic ecologist with the University of California at Davis. 'These invasive species get us further and further away from the pristine condition of the lake that people would like to see.' … Scientists worry that the clams will consume so much plankton native fish will starve." (UPI, 2009) Ironically, few noticed that at the same time UNR researchers were blaming Asian clams for causing algae, TERC researchers were claiming that fish might starve because the clams ate too much algae.

Reuter was one of the UC Davis and UNR coauthors of the December 2008, *Asian clam of Lake Tahoe: Preliminary scientific findings in support of a management plan*. The report says, "Ecosystem scale impacts to zooplankton[6], [algae], and fish are likely if current [Asian clam] populations expand their range to become the lake-wide nearshore dominant species." Not stating whether the negative impacts were from nurturing algae or eating too much of it, the report concludes: "Given these potential impacts, there is increasing interest in developing an effective control strategy for clam populations, predicting their spread, and preventing other future invertebrate introductions (e.g. quagga and zebra mussel, New Zealand mud snail, [and] spiny water flea) in Lake Tahoe" (Wittmann, et al., 2008).

Scientists seldom single out their colleagues' errors, but instead simply state the erroneous premise, give the evidence for opposing conclusions, and allow researchers and academics to sort out what is correct. That is why it was startling in June 2014 when Cornell University posted on their *New York Invasive Species Clearinghouse* website: "Recent justifications for expensive and intensive management actions (e.g., Lake Tahoe's Asian Clam Response) overstate evidence in the scientific literature. For example, in *Asian clam of Lake Tahoe: Preliminary scientific findings in support of a management plan*" states: 'Asian clams are known to aggressively out-compete native invertebrate communities.'

"Here the report's authors incorrectly interpret and cite the findings of Karatayev et al. (2003). Karatayev and colleagues studied [Asian clams] in a Texas reservoir. Although the clams dominated the total animal biomass of the reservoir sediments (up to 95%), [they were] not associated with declines in native

---

[6] Small aquatic animals

biodiversity. In fact, [the clams were] found to co-occur with an abundant population of native unionid mussels.

"Studies claiming that the Asian clams have an impact on native bivalves (particularly unionids) are often anecdotal and only report the spatial distribution of bivalves after invasion (Strayer 1999). They assume that non-overlapping distributions of [Asian clams] and native mussels indicate that [Asian clams] have out-competed the native species. As Strayer points out, this is just one possible explanation. [Asian clams] could also prefer different habitat than native mussels (e.g. sandy vs. silt/gravel). ... It is difficult to tease apart these changes versus direct impacts of invading [clams] that also happen to do well." (Cornell, 2014)

Because Wittmann, and Chandra didn't publically retract their initial premises or acknowledge their errors, few were perplexed or even learned that six years after Asian clams were maligned at Tahoe as biodiversity destroying invasives that the Tahoe research academics, who led the campaign to denigrate the clams, were wrong; that they had prejudiced their results without presenting all probable possibilities; and that they had misstated facts from scholarly studies to justify their findings.

Prior to releasing their preliminary report, the researchers knew that Lake Tahoe's Asian clams were stunted by harsh environmental conditions and had only heavily colonized patchy beds in the shallow warm water on the lake's south shore. Since then they have acknowledged that the extent of Lake Tahoe habitat favorable for Asian clam colonization is miniscule compared to what they initially projected—from potentially 12,000 acres to less than 500 acres of Lake Tahoe's nearshore—and that Asian clam infestations have spread to few other areas.

— 3 —

The Winds of War

The rigid attitudes and militaristic metaphors that characterize the debate about exotic species make for poor science and policy-making.

~ Mark Davis, Biology Chairman, Macalester College

The most passionate calls for the prevention and control of non-native species are heavy with metaphors of war. The same month TRPA said Asian clams could facilitate the introduction of dreissenid mussels, a Zephyr Cove resident wrote in a *North Lake Tahoe Bonanza* op-ed: "We are going to lose the zebra mussel war. The use of the term 'war' is not an exaggeration. The zebra mussel is a terrorist in every sense of the word. It is an invader that has the full intention of taking over and becoming victorious by destroying everything that gets in its way. Its weapons are truly weapons of mass destruction and its numbers are astronomical. Once it gets a stronghold in enemy territory, nothing will stop it from multiplying and continuing its rampage of destruction."

Five months later, a headline in the *Tahoe Daily Tribune* blazed: "War Declared on Asian Clams at Lake Tahoe." The article began, "Scientists are preparing to wage an all-out war against another threat to Lake Tahoe's famed pure waters: Asian clams. The clams have turned up in numerous locations along Tahoe's southeast shore and prompted concern that they could pave the way for even more destructive invasive species such as quagga."

Alarmist public rhetoric was understandable, but the claims and misstatements made by TRPA and their research academics that led to the near hysteria were irresponsible. A year earlier, Thomas Whittier had led researchers from Oregon State University in a study for the U.S. Environmental Protection Agency that analyzed metadata from the EPA's Monitoring & Assessment Program and the Geological Survey's National Stream Accounting Network. "A calcium-based invasion risk assessment for zebra and quagga mussels" was published in *Frontiers in Ecology and the Environment*, the Ecological Society of America's peer-reviewed journal. It showed unequivocally that dreissenidae are unable to propagate in water with calcium concentrations as low as Tahoe's.

The 2008 study evaluated 15 major dreissenid mussel studies and correlated 5 to 10 years of calcium data from 3,091 U.S. water bodies. The study established calcium requirements for quagga and zebra mussel survival, colonization, and infestation. Using the concentration of dissolved calcium in each of the water bodies, the U.S was mapped for the risk of infestation by the mussels and cross-checked against actual field observations. Eighty-two ecoregions were categorized into four ranges of risk of potential invasion: mussel free (Very Low risk—9.4% of the total U.S. land area); non-reproducing colonies (Low risk—11.3% land area); persistent, potentially harmful colonies that would very probably require control measures to prevent infrastructure damage (High risk—58.9% land area). The remaining 19.8 percent of the country was "Highly Variable." (Whittier, 2008)

As TERC researcher Marion Wittmann accurately stated, quagga mussels require calcium concentrations greater than 12 ppm for short-term survival, and the Whittier study categorized ecoregions falling below that threshold as being risk free. Two other thresholds were found to be at 20 and 28 ppm. Patchy, non-harmful sink populations, sometimes developed when calcium was greater than 12 ppm but was less than 20 ppm, and a "Low" probability of colonization was established for that range. Between 20 and 28 ppm, colonization was Highly Variable, and other environmental factors appeared to affect populations. In water with dissolved calcium greater than 28 ppm, persistent reproducing and potentially harmful colonies formed, and they were designated as "High" risk.

In the nine years following Whittier's study, the thresholds proved to be highly reliable and no harmful infestations, requiring control, occurred in the 20 percent of the country classified as Very Low or Low risk. Although comprising only 9 percent of the country, four U.S. regions have minimal limestone deposits, the primary source for calcium in lakes and streams. They are the south-central Lower Mississippi Valley, Appalachia, the Pacific Northwest, and the Sierra Nevada. These areas comprise most of the Very Low risk region mapped for EPA, and all are mussel-free.

Because successful introduction by a non-native aquatic species depends on its ability to survive under the full range of seasonal temperature and environmental conditions it will endure in a new habitat, in addition to analyzing biological requirements, rigorous water quality analysis and scrutiny of seasonal weather conditions are critical to assessing the risk of invasion. After stringent analysis of invasion potential, a minimum of five years of data where colonization has occurred in similar environments should be compared before assessing risk. Even if conditions are similar, it is not uncommon for quagga to rapidly colonize, as they did in north Texas and Oklahoma reservoir lakes, and then have their populations crash and fall to harmless levels.

At Tahoe, a comprehensive risk assessment for quagga and zebra mussels isn't actually needed. The lake's dissolved calcium averages 9.16 ppm, well below the threshold required for dreissenidae survival and less than half of that needed to sustain harmful colonization. Gerald L. Mackie and Renata Claudi, two of the world's foremost aquatic biologists, state that quagga can't survive in lakes with calcium less than 10 ppm or propagate in lakes with calcium less than 15 ppm—all of the major High Sierra lakes have calcium levels less than Tahoe's and are risk free. The Whittier study established the calcium thresholds for determining the risk of zebra and quagga mussel survival and infestation, and established the guidelines used by the nation's 40-plus AIS management programs to assess the threat of potential mussel infestations in their regions. One AIS management program—Lake Tahoe's—ignored the Whittier study and did its own.

With the objective of making a "prediction of invasion into Western water bodies," six UNR and UC Davis researchers, led by Sudeep Chandra, conducted a quagga survivability experiment for TRPA between the 2008 and 2009 boating seasons. Funded by a $20,000 grant, the UNR Aquatic Ecosystems Laboratory tested eight adult quagga mussels from Lake Mead for their ability to survive and remain fertile for 51 days in water from the Tahoe Keys. The overriding aspect of the study was that its parameters were flawed: it takes a full year for female quagga to mature. To be meaningful, test conditions comparative to the habitat being studied must be simulated for the year from veliger stage to reproductive maturity. Then if the mussels survive, comparisons should be made to infestations occurring in comparable habitats. No aspect of that common process was adhered to.

The chemical makeup of Western water bodies is as varied as the West's weather and geology. Lake Mead is surrounded by alkaline desert and has dramatically different chemical properties than Lake Tahoe's water, where the lake bottom is igneous rock. To take adult mussels from a hospitable habitat with exceptionally high calcium and test them to see if they can survive for a short time in a lab in low-calcium water has no scientific value.

Putting into perspective the insufficient funding provided to examine whether quagga could survive and reproduce in Lake Tahoe, a $321,000 grant had been awarded earlier in the year by the U.S. Forest Service to UNR and TERC to examine the potential of Asian clams "to alter the ecology of the lake"—16 times more than was allocated to study if quagga mussels, the presumed greatest AIS threat to Tahoe, could survive there.

Possibly because the quagga survivability experiment was redundant to the federally-funded Whittier study, it was not funded by usual sources. In 2003 the Southern Nevada Public Land Management Act was amended to allow the Bureau of Land Management to direct $38 million a year from public land sales to the Tahoe Environmental Improvement Program. The EIP became the primary funding source for environmental studies and programs in the Tahoe area, and grants from the federal program funded 362 projects, including most of Tahoe's AIS studies. The $20,000 experiment was funded by the Army Corps of Engineers.

Quagga mussels for the experiment were transported to the UNR aquatics lab from Lake Mead and placed in water from the Tahoe Keys Marina in South Lake Tahoe. Keys water is the most nutrient and calcium-rich water at Tahoe. The experiment used water with 13.5 ppm calcium—a third higher than the lake's average nearshore calcium level, higher even than the average calcium level in the Tahoe Keys, but was 12.5 percent higher than the survival level for quagga.

Even though the quagga were rapidly losing weight by the end of the 51-day experiment, remarkably, the study predicted: "The possibility exists for at least adult quagga to survive, grow, and reproduce in the Tahoe environment; the diligent monitoring of recreation vehicles putting into western lakes-of-interest is prudent. The assumption that western oligotrophic water bodies low in calcium are at very low to low risk of quagga invasion is not necessarily supported. ... We recommend continued monitoring and prevention efforts."

The researchers didn't conclude that mussels could survive in Tahoe, only that it was "unclear" whether the lake's habitat supported quagga establishment. However in the study's report, Whittier's *A calcium-based invasion risk assessment for zebra and quagga mussels* was dismissed "because quagga appear to have different environmental tolerances than zebra mussels. ... The potential risk of invasion of western water bodies may be underestimated by using zebra mussel-based risk assessments"— a claim refuted by several peer-reviewed studies that report that quagga have a higher calcium requirement than zebra mussels. The *Quagga Mussel Risk Assessment* wasn't peer-reviewed and was released nine months after TRPA passed mandatory boat inspections and a year after the Whittier study had conclusively shown that dreissenid mussels can't infest Lake Tahoe.

After reviewing the science and the *Quagga Mussel Risk Assessment*, journalist David Bunker wrote in *Moonshine Ink* in May 2013: "For years, Tahoe residents heard little about the science that showed Lake Tahoe and Donner Lake were at very low risk of mussel invasion. Instead, they heard stories about Lake Mead, with photos of quagga-encrusted boat propellers from the Nevada reservoir that is a starkly different environment than the Sierra Nevada, and stories about the millions of dollars of damage they create. And when the public did hear about scientific studies, they heard mostly about one 51-day, low-budget study that has heavily influenced Tahoe's AIS policy.

"The report did not research the full life cycle of quagga mussels, including the ability of vulnerable mussel offspring (called veligers) to survive in Lake Tahoe. The ability of mussels to reproduce in a [habitat] is the true measure of risk, since the threat of mollusks is directly tied to their rapid colonization of a habitat. A non-reproducing 'sink population' of mussels would only survive for the animals' three to five-year lifespan." Praising his research, in 2014 Bunker's 1,500 word article on quagga's inability to propagate in Lake Tahoe was awarded first place for best environmental story by the National Newspaper Association.

A decade before quagga were found in Lake Mead, sink populations of zebra mussels appeared in New York State's Lake George, an oligotrophic lake whose average calcium is slightly above 12 ppm. More than 10,000 boats annually launch from the lake's 30 public ramps and provide a constant source of zebra mussel introductions. In the decade after zebra mussels were first detected, nine small sink colonies were found at locations spanning the length of the lake, but no harm was attributed to them, and in the six years after 2009, no zebra mussel colonies were detected in Lake George.

Lake George's experience is used to show that dreissenid mussels can survive in an oligotrophic habitat where calcium is as low as 12 ppm. This conclusion ignores the fact that none of the colonies persisted. It also overlooks an experiment by the Darrin Fresh Water Institute of Rensselaer Polytechnic Institute that placed healthy zebra mussel veligers up to two weeks of age in Lake George water. DFWI Director Sandra Nierzwicki-Bauer, Rensselaer biology department chair reported, "Older mussels have no difficulty developing when placed in Lake George water, but when larvae less than two weeks old are placed in the lake's water they all die within seven days." (Rensselaer, 1998)

During their larval stage, mussel veligers drift with the currents for three to five weeks, filtering calcium and feeding by hair-like cilia. The reason mussel veligers quickly die in very low-calcium water is because calcium is the primary material in their shells, and unless pH is exceptionally high, during their high-growth development stage they require water with calcium greater than 15 ppm. In the exceptionally cold, low-calcium waters found in the High Sierra, mussel veligers die of calcium deprivation within a few days. (Hincks & Mackie, 2011)

Lake George's calcium is a third higher than Lake Tahoe's, and Chandra ignored Rensselaer's findings that veligers are more calcium dependent than adult quagga and tested mature mussels. When he released the *Quagga Mussel Risk Assessment*, he wrote, "It appears adults can survive [in Lake Tahoe]. The hypothesis we were testing is there would be no survival. I was quite surprised." If that were true, it's more surprising that he didn't recommend to TRPA's Board of Governors that they delay mandatory inspections until he at least completed his experiment, which was months before the start of the 2009 boating season.

If Chandra's hypothesis that the quagga wouldn't survive had been confirmed—and apparently if the experiment had been extended much longer, all would have died—boat inspections wouldn't be necessary. But Chandra wasn't surprised. He knew before conducting the experiment that all of the mussels should make it because they had sufficient calcium reserves in their shells to survive 51 days in any water. He also knew what other nutrients were needed for their survival.

The *Quagga Mussel Risk Assessment* states: "Dietz et al. (1994) found that zebra mussels survived at least 51 days in water containing minimal concentrations of potassium, sodium chloride, and magnesium but no calcium and suggested that the mussels survived by mobilizing calcium from [their] shells to maintain critical levels of blood calcium for muscle function." Chandra tested for no unknown science and avoided even addressing his primary objective: to determine the survival, growth, and reproduction potential of quagga in Lake Tahoe. That science was also as well-known as the fact that the mussels would survive for

51 days, and competent muscologists knew then that quagga couldn't propagate with calcium levels less than 15 ppm.

The *Quagga Mussel Risk Assessment* closely replicated the parameters established in Dietz' *Osmoregulation in Dreissena polymorpha*, done by the Department of Zoology and Physiology at Louisiana State University. The 15-year-old study was done during the flurry of experimentation and analysis when scientists were determining effective methods of controlling zebra mussels in the Great Lakes and Mississippi Valley. It tested zebra mussels for essential minerals critical to their survival, and reported that "by mobilizing calcium from their internal stores" to maintain blood calcium, adult dreissenidae are able to survive more than 51 days in pure, calcium-free water, but can't survive more than 30 days in water that doesn't have minimal concentrations of magnesium and other common salts.

The LSU study noted that unlike many freshwater animals that can survive for more than 30 days in pure water, dreissenidae can't. They were the most salt-ion dependent bivalves the LSU team researched. The study also pointed out that lakes low in calcium are usually low in magnesium, and that unlike calcium, magnesium isn't stored in shell material. For adult dreissenidae to survive beyond 30 days, magnesium is more critical than calcium. (Dietz, 1994) Other than calcium thresholds, there is a paucity of data on concentration thresholds for other minerals for dreissenid survival. It would have been more beneficial for Chandra to test for minimum concentrations of magnesium needed during the mussels' lifecycles and its availability in Sierra lakes—calcium requirements for all dreissenid life stages are well established.

The same year that the Chandra-led risk assessment was published, Canadian biologists Gerald L. Mackie and Renata Claudi released the Second Edition of *Monitoring and Control of Macrofouling Mollusks in Fresh Water Systems*. The book is the definitive text on dreissenid biology; physiological tolerances and requirements; dispersal potential; and monitoring and control methods. The author of four books, 12 text chapters, and more than 150 peer-reviewed scientific articles, Mackie is considered the world's preeminent authority on zebra and quagga mussels. Claudi is Chief Scientist for RNT Consulting Inc., North America's leading consulting firm for risk evaluation for invasive mollusks and appropriate dreissenid mussel control methods.

A compendium of dozens of peer-reviewed studies and dreissenid research, the book details the biological and physical requirements of zebra and quagga mussels during their life cycle, and is the primary authority on assessing risk and control of invasive mollusks. Explaining how phosphorus and nitrogen drive chlorophyll *a* levels that in turn affects Secchi depth, the scientists developed the most comprehensive guide for determining the risk of quagga and zebra mussel invasion. The following table summarizes primary thresholds for dreissenid mussels during their two principal life stages:

Table A: Quagga Survival Thresholds for Adults and Veligers

Parameter	Adults do not survive long-term	Uncertainty of veliger survival	Lake Tahoe levels
Calcium (ppm)	< 10	< 15	9.16*
Mean Summer Temp (°F)	< 64	< 64-68	< 64**
Secchi Depth (ft.)	> 30	> 10	> 65**
Chlorophyll *a* (µg/L)	< 2.5	< 2.0-2.5	< 1.1**
Phosphorous (µg/L)	< 5	< 5-10	< 3**

* *Quagga Mussel Risk Assessment* (Mackie & Claudi, 2009)
** *Tahoe: State of the Lake Report 2015*

Measurements of Lake Tahoe's water quality parameters are outside all five of the survival thresholds for dreissenids. As demonstrated at Lake George and hundreds of other lakes, if any of the key parameters doesn't meet a survival threshold, a sink population of mussels may form, but they won't persist. Tahoe's extremely low-calcium, ultra-oligotrophic water assures that even a massive introduction of zebra and quagga mussels would die off within a couple of months.

Four years after completing the *Quagga Mussel Risk Assessment,* Chandra defended his study. Saying "seven of eight adult quagga survived 51 days [and] that is an indication to us that they are able to survive in the short term," his experiment had indeed reconfirmed that quagga can survive for a short time in low-calcium water—just as they can in calcium-free water.

The first phase of the *Quagga Mussel Risk Assessment* was a baseline study to establish Tahoe's average nearshore calcium. Water samples were taken in May and November 2008 from 83 locations spanning Lake Tahoe's entire 76-mile shoreline. Chandra's survey included 23 high-calcium locations that were near culverts, runoff outlets, boat launches, or marinas where water is subject to calcium leaching from concrete, and included readings from the Tahoe Keys, a marina-community behind the lake's shoreline. Consequently, the average calcium concentration of 9.16 ppm likely represents the highest probable calcium baseline for Lake Tahoe's nearshore—the region in all lakes where nutrients are greatest. This observation is supported by the fact that Lake Tahoe has the highest calcium level of any large High Sierra lake.

The most glaring discrepancy in Chandra's research is that he contradicts his own findings and misstates baseline values that he established. His first-phase research determined that the lake's average nearshore calcium is 9.16 ppm, but in later studies, including his 2013 *Inventory of aquatic invasive species and water quality in lakes in the Lower Truckee River Region* he wrote: "In contrast to the Whittier assessment, we also utilized [the *Quagga Mussel Risk Assessment*] based on Tahoe water, which contains approximately 13 ppm calcium"—grossly misstating the dissolved calcium data determined by his own study.

It is unconscionable for a researcher to misrepresent their work and distort critical statistical values and data that their studies investigate and that guide important public policy, but that is what Chandra does. His laboratory experiments on quagga consistently use test water with dissolved calcium artificially raised above 13 ppm—40 percent greater than the baseline level that he determined. And instead of using water in his tests with quality equivalent to Lake Tahoe's, he uses polluted, nutrient-rich water from the Tahoe Keys' manmade ecosystem.

In the *Quagga Mussel Risk Assessment*, Chandra wrote: "After a 51 day exposure, quagga adults survive, exhibit positive growth [they were rapidly losing weight when the experiment ended], and have the potential to release gametes. This study did not have the funding to follow the reproductive cycle of the mussels to determine if they could produce mature veligers." At the time of the 2009 report, dozens of peer-reviewed studies had conclusively demonstrated that quagga and zebra mussels could not reproduce in Lake Tahoe quality water and couldn't survive in the lake for more than a couple of months.

By misstating data, exaggerating results, using scare tactics to create unfounded fear of the damage quagga could wreak, and promoting an overtly biased experiment to cast doubt on whether quagga could propagate in Lake Tahoe, a hastily enacted boat inspection program was given credibility. And because Chandra claimed Asian clams created calcium "hot spots" and were a facilitator for quagga mussels, untold numbers of the lake's native aquatic animals became collateral damage in TRPA's multimillion-dollar war on Asian clams—an expensive war the agency lost.

— 4 —

Scientific Ambiguity or Deceit

It's unpredictable. This is the entire problem with biological invasions. If mussels establish in Lake Tahoe, then the worst-case scenarios are what our colleagues have observed back east, where they have to replace water pipes, treat the sewer systems, and things like that. ~ Sudeep Chandra, Limnology Professor, UNR

Biological invasions are predictable. Modern land, sea, and air transport make it almost certain that anywhere a plant or animal species can find a hospitable habitat, it will take up residence. Despite conservationists' efforts to restrict species to their native range, even the earth's largest animals cause improbable invasions. It is difficult to imagine that South America could be invaded by hippopotamuses, Africa's most dangerous mammal, but it happened. In the 1980's, drug czar Pablo Escobar stocked his Columbian estate's private zoo with a pair of hippos. When he was killed in 1993, nobody wanted them, and they were released into the Columbian jungle, a habitat very much to their liking. Now there are about 60 African hippos in Puerto Triunfo.

National Public Radio reported in May 2014: "The locals just want them out because they know they're dangerous. Nobody wants to mess with them, and it's a big question about what to do with them. The locals are looking for a solution to their hippo infestation, but finding a refuge that can take a large population will be difficult. Nobody wants to take them because they are expensive to feed, to keep, and very difficult to control." In comparison, Tahoe's Asian clams seem to be a minor issue.

Columbia's invasive hippos underscore a conundrum in controlling nonnatives: The higher up the food chain an animal is, the more protected it becomes. Hippos are destructive; however if hunters were allowed an open season on alien South American hippos, animal rights advocates would rise up in force. Few are bothered by the killing of millions of Tahoe's native mollusks or flatworms, but most are passionate about defending the lake from quagga mussels—an unfairly maligned species that can be a water quality and ecosystem asset.

S.A. Ostroumov studied zebra and quagga mussels in their native Ukraine habitats. In a peer-reviewed study published in *Aquatic Biodiversity*, he concludes that mussels and other filter-feeders play pivotal ecosystem roles, including: "(1) ecological repair of water quality, (2) contributing to reliability and stability of the functioning of the ecosystem, (3) contributing to creation of habitat heterogeneity, [and] (4) contributing to acceleration of [elimination of polluting] chemical elements." Unfortunately Lake Tahoe won't realize those benefits—quagga can't survive there.

When the *Quagga Mussel Risk Assessment* was released Sudeep Chandra and Marion Wittmann said researchers should examine survivability of mussels in their larval stage, as soon as funding for more studies was secured. Eighteen months later in October 2010, Chandra and Kumud Acharya of the Desert Research Institute's Las Vegas center were awarded a $338,000 grant from the U.S. Forest Service as co-principals for a quagga survivability follow-up study at Lake Tahoe. (Five years later, when asked when the study's results would be available, Acharya said that all questions should be directed to Chandra, the principal investigator.) The Nevada professors' grant proposal noted that "survival from veliger stage to adult stage is key in the establishment of mussels in any water—this indicates the ability for species establishment after introduction."

The researchers' observation is correct, but many studies, including Rensselaer's veliger tests, had previously demonstrated that quagga veligers don't mature into juveniles in water that is as cold and pH neutral as Tahoe's, and has calcium less than 15 ppm. Because dreissenid veligers don't have shells to store calcium reserves, the juvenile mussels are more dependent on dissolved calcium than adult mussels. Table A on page 56 shows the other minimum environmental thresholds for quagga veligers.

Chandra and Acharya dismissed accepted science and peer-reviewed research and wrote: "Whether juveniles, and thus a reproducing population can sustain is unknown. [Our] research will directly assess the habitat suitability of Lake Tahoe and its watershed to support the establishment of quagga mussels by testing the impacts of Tahoe conditions on the survivability of veligers to sub-adult stage."

A year after the grant was awarded, *Vegas Seven* reported: "It's unclear whether [quagga] would take to Lake Tahoe as they have to Lake Mead. For one thing, they need calcium to grow shells and eventually spawn—and, while there's plenty of calcium in Lake Mead, there's not nearly as much in Lake Tahoe. The colder water temperatures at Tahoe's 6,225-foot elevation may also impede quagga growth.

"'This is the reason we're trying to do some of these tests now,' says Chandra of scientific studies that began in November, with the first results to be available by July [2012]. 'Can these things survive, and how fast would they grow if they did survive? And at what stages would they be limited or not?'"

Results weren't released by July. In fact, no final results or reports from any of Chandra's follow-up quagga experiments were released in the seven years since the 2009 *Quagga Mussel Risk Assessment*—long after the public had accepted that quagga mussels were a major threat, and Tahoe agencies had proclaimed that watercraft inspections effectively prevented their spread.

Published a year before Chandra's quagga study, Mackie and Claudi's, *Monitoring and Control of Macrofouling Mollusks in Fresh Water Systems* had conclusively answered the questions that Chandra claimed were unresolved. Seemingly, Chandra had to be aware of the highly-acclaimed text that is the authoritative reference on invasive mollusks, and he was. At a presentation in Truckee, he was asked if he was familiar with Renata Claudi and RNT Consulting's work for the California Department of Water Resources. He said yes and changed the conversation to a study Claudi did on Upper Colorado River reservoirs where patches of quagga had been found in "low calcium" water. As at Lake George, the Colorado mussels were sink colonies that had formed where concrete had leached near dams, but did not propagate.

Chandra is required to file annual progress reports on an active project with the Bureau of Land Management, his funding agency. Reports on quagga survivability were filed for FY 2013 and 2014. In the first, he reported: "Veligers were collected from Lake Mead and exposed to varying water treatments: two replicates of Lake Mead (control), two replicates of Lake Tahoe (Tahoe Keys), and one calcium amended treatment (Keys-25 ppm). Treatments were kept for 28 days at room temperature. Results for the survivability and growth of Lake Mead and Tahoe comparative treatments indicate 24% veliger survivability."

In the *Quagga Mussel Risk Assessment*, he had written that "a range of finer-scale calcium levels" needed to be tested. It was only logical that the calcium concentrations would be scaled down from the known survival threshold of 12 ppm to Tahoe's nearshore average of 9.16—not up from a low test sample of 13.5 ppm, as he did. Chandra's 2009 report states that in future tests, mussels should have "enough food for maximum growth potential." He apparently justified lacing the test water with calcium to maximize the mussels' growth. Even though the 2013 tests were absurdly unrealistic and deviated from known control thresholds, they didn't produce the results Chandra desired, so he did them again, using even greater calcium concentrations.

From the 2013 interim report it is impossible to determine what "comparative" Lake Mead and Lake Tahoe treatments are—the water chemistry and ecology of the two lakes are disparately dissimilar, and survivability over 90 days is not the issue. Tahoe's dramatic climatic and environmental changes during the year-long maturation cycle when veligers develop into juveniles, settle to the bottom, and mature into adults is impossible to simulate in a laboratory. It is meaningless to test the survivability of mature adults for 90 days and veligers for 28 days under independent and artificially controlled lab conditions to determine reproductive sustainability. Regardless, a major control discrepancy invalidated the veliger tests: no Lake Tahoe water has ever reached room temperature for 28 days.

Lake Tahoe has long been the world's most closely monitored lake. UC Davis researchers have recorded a wide array of Tahoe's environmental conditions since 1968. And an annual "State of the Lake" report has been produced since 2007. The report graphically details water quality changes and trends in the lake's physical and biological properties, clarity, temperature, and nutrient levels. The 2015 report explains that the lake's surface water is warmest in July, and in the 16 years after 1998 the average July surface temperature was 64.7°F. Veligers die if water temperature doesn't average at least 64°F for the month they free-float, a thermal condition that seldom occurs at Lake Tahoe.

Baseline controls in Chandra's study were egregiously violated and the water used in the experiment wasn't from Lake Tahoe. It was taken from the Tahoe Keys Marina, a dredged lagoon that sits behind the lake's shoreline and is fed by the Upper Truckee River. Tahoe Keys' water was transported to Las Vegas where it was filtered to remove all plankton and foreign matter and was stored in 5-gallon buckets. The test water pH in the veliger study rose above 8.0, a level much higher than Tahoe's near-neutral water (7.0 ±0.2), and well above 7.5, a level where veligers are benefited by elevated pH (Claudi, 2011). The tests were a total sham—environmental conditions weren't replicated, and as Charles Goldman had demonstrated 40 years earlier: fine-filtering water doesn't remove dissolved nutrients—a fact taught in high school chemistry classes, but ignored by the researchers.

A second objective of the study was to determine if Lake Tahoe could support "adult [quagga] survival, growth, and the potential to produce viable gametes." Chandra reported: "Adult mussels were collected from Lake Mead and subjected to experimental Lake Tahoe water treatments. Six treatments were evaluated that encompassed natural waters from two Tahoe locations (Cave Rock, Tahoe Keys), three calcium [supplemented] Tahoe treatments (Keys-20 ppm, Keys-25 ppm, Keys-32 ppm), and a control (Mead). Twenty [mussels] for each treatment were held for 90 days at [68°F]"—the minimum temperature needed to sustain the mussels growth, but 5 degrees warmer than Lake Tahoe's summer average. "Survivorship over 90 days was high for Tahoe Keys water [supplemented] with calcium (80 to 90%) and comparable to the control, Mead water (95%). In contrast, low survivorship was exhibited by natural Tahoe water treatments, Keys (0%) and Cave Rock (35%)." The results produced what would be normally expected and demonstrated no deviation from the known science.

The report cites an issue that arose: "Interpretation of the adult mussel experiment is complicated by the abrupt die-off of mussels subjected to the Tahoe Keys water treatment. From day 8 to day 20 of the experiment, 19 of 20 mussels died in the Keys treatment. The result is difficult to interpret given that the source water was the same for the Keys calcium amendments. ... Due to this complication, we plan to replicate the experiment with Tahoe Keys water and several calcium addition treatments (Keys-16 ppm, Keys-23 ppm) to verify results."

Chandra's adult mussel experiments demonstrated exactly what muscologists had known for more than 20 years, but the experiments didn't provide the results that Chandra wanted, so he changed the experiments' parameters and tried again to nurse the adult mussels through 90 days in his lab. The 2013 report to the Forest Service concludes: "Another experiment is needed to help guide construction of the risk assessment. ... A second adult experiment will be conducted to evaluate survival, growth, and reproduction. Adult mussels will be collected from Lake Mead and held in Tahoe Keys water in the laboratory for 90 days. This second experiment will be completed during the fall-winter 2013-2014 or spring 2014. Completion of this second experiment will allow us to move forward with assessing the probability of risk."

The second attempt to sustain adult quagga for 90 days in the lab didn't "replicate" the first attempt, and Chandra reported that "survivorship over 91 days was high for Tahoe Keys water (80%) and comparable to Tahoe Keys water [supplemented] with calcium (90 to 95%)." More ambiguous than the 2013 annual report, the 2014 report says that water temperature averaged 68°F and the report didn't provide the average pH for the three-month experiment, but it was the results from the veliger study that was done at the Desert Research Institute's Las Vegas campus that were truly groundbreaking. They indicated that the researchers achieved a scientific breakthrough. Chandra wrote: "Veligers raised in Lake Tahoe water … survived and grew at an equal growth rate [to those in Lake Mead water].

Even though tests had extended three years beyond their promised completion, had been completed for months, and had produced apparently stunning results, Chandra again postponed making an evaluation. He wrote: An assessment "using the [test] results has not begun. Information from the experiments will be used to construct the risk assessment"—a surprisingly odd deferment, considering his unprecedented findings.

A month after the Forest Service posted Chandra's 2014 annual report, an early draft of this book was sent to each TRPA Governing Board member, key TRPA managers, Chandra and several of his colleagues. An accompanying cover letter to the TRPA governors noted that this book wouldn't be published until Chandra's final report was released so that the results of his four-year quagga study could be evaluated. And because a release date for a risk assessment had not been given, the governors were asked for their assistance to have him release his findings to date.

At the time, TRPA was lobbying the Nevada Legislature for funding to match California's two-year commitment to underwrite the agency's funding shortfall for boat inspections. If Chandra had concluded that quagga couldn't propagate in Lake Tahoe, Nevada State funding for the boat inspection program may not have been granted. Instead of releasing his experiments' results, Chandra muted the discussion about the necessity of the inspection program by announcing to the *Reno Gazette-Journal* that "adult quagga can live in Tahoe's water and that successful reproduction can occur and juveniles survive to repeat the cycle as adults." His eight years of research had shown no such thing.

Headlined: "Study suggests quagga mussels could thrive at Tahoe," the front-page article went on to report: "Quagga could survive and reproduce in Lake Tahoe's waters, particularly in places where a different aquatic invader is already present and where algae are flourishing in shallow waters.

"Research confirms earlier work that indicates adult quagga mussels can live in Lake Tahoe water and that successful reproduction can occur and juvenile mussels survive to repeat the cycle as adults. Experts had hoped low calcium levels in the alpine lake might help protect Tahoe from major infestations of the type in Lake Mead and other lakes. It appears that's not the case."

Because the temperature, pH, and calcium in the water used in the veliger experiments were unnaturally elevated above survival thresholds, it wasn't surprising that veligers survived in the tests. But Chandra's claim to the *RGJ* that they "did so long enough to attach to rocks and other surfaces under natural conditions, after which they [grew] very quickly into reproducing adults" was a complete fabrication—the experiments didn't test for those factors, and they simply couldn't happen in less than a year, the time it takes for female quagga to reach reproductive maturity. The veliger component of the study hadn't investigated veliger maturation, only survival, and no veliger was sustained beyond 28 days, much less had quickly grown into a reproducing adult, but that is what was communicated to the public and the agencies that support and rely on Chandra's work.

Chandra also took the opportunity to repeat his bogus theory that Asian clams increased Tahoe's dissolved calcium to explain how his experiments were able to defy accepted science. Noting that "survival rates climbed to more than 90 percent when calcium levels were raised to match those measured in water near beds of invading Asian clams, which now cover the lake bottom across much of the southern portion of Lake Tahoe," Chandra said, "'It's like giving them a smorgasbord. It clearly points out the risk of Asian clams facilitating (quagga) survival.'"

The *Reno Gazette-Journal* front-page article relieved pressure on TRPA to reexamine the basis for the Tahoe watercraft inspection program. The *RGJ* article reported: "Latest research indicates that not only Lake Tahoe but other bodies of water experts thought might be immune from the quagga threat could instead support 'sustainable quagga mussel populations. Research clearly demonstrates a need to continue screening boats entering Lake Tahoe in hope to prevent quaggas from being introduced.'"

Newspapers throughout the region reprinted the *RGJ* story, and using only newspaper accounts, Casey Beyer, chairman of the TRPA governing board, strengthened Chandra's claims and highlighted quagga mussels as Tahoe's primary AIS threat when he wrote in an op-ed press release: "A new scientific study highlights the importance of collaborative work to prevent the introduction of aquatic invasive species at Lake Tahoe, and confirms what many stakeholders have feared: That the invasive quagga mussel, if introduced, could establish in our mountain lake's clear, iconic waters.

"The study by UNR and Desert Research Institute found quagga mussels could survive in Lake Tahoe, particularly in areas with elevated calcium levels from already-established Asian clam populations and in areas of higher algae growth, as found in some nearshore parts of the lake. ... Why are quagga a concern? They are known as prolific breeders and can form sprawling colonies that carpet rocks, lake bottoms, boats, and other structures, littering beaches with shells and causing major environmental, economic, scenic, and recreation impacts." Although greatly exaggerated, except for Asian clams elevating dissolved calcium, Beyer's claims are true—only not at Lake Tahoe or any other lake in the Sierra Nevada range.

Most Tahoe residents already believed that quagga were a serious threat to the lake, and the articles describing Chandra's new conclusions caused little local stir. The Nevada Assembly approved TRPA's funding request for boat inspections, and the totally unnecessary and costly inspection program continues.

If Chandra's tests had been realistic and honest, for muscologists, biologists, and AIS managers his findings would be groundbreaking. But because he didn't publish a report, release data, or submit his findings to a scientific journal for publishing, the integrity and accuracy of his extraordinary claims haven't been evaluated and have gone unnoticed in the scientific community—where, if they had been promoted, they would have been quickly dismissed as fraudulent. The science is settled on the environmental and biological requirements of dreissenid mussels; the issue is of only regional interest (Lake Tahoe and Donner Lake are the world's only lakes that have mandatory boat inspections and calcium below 12ppm); and in the spectrum of public policy issues, the necessity of boat inspections is a very low priority.

The March 2015 *Reno Gazette-Journal* also reported: "'We were surprised because we felt [the mussels] would need a lot more calcium,' said Kumud Acharya, a scientist with the Desert Research Institute's Las Vegas center who headed up the veliger component of the study. 'Apparently there's enough.'"

A search of scientific papers presented to the University of Nevada, Las Vegas revealed that the research for the veliger component of the study was done by a UNLV graduate student, for her Master of Science thesis in Water Resources Management. The student also managed the DRI Quagga Mussel Research Lab, and her duties included the supervision of the sample water storage tank and specimen aquariums for the veliger survivability study. Her research didn't confirm Acharya's assessment that Lake Tahoe has sufficient calcium to sustain quagga, and her thesis highlights the extremes Chandra goes to in order to validate false conclusions from his previous experiments, rationalize unfounded theories, and justify his questionable recommendations.

Having the veliger research done on the UNLV campus was logistically logical and using a co-investigator gave the study the appearance of impartiality—but it was highly prejudiced. Chandra acknowledges that calcium is the critical environmental variable for quagga establishment and survivability, and that water pH and temperature affect quagga's ability to absorb dissolved calcium. However, the student was given unrealistic control levels for her tests, her results are biased, and her conclusions are prejudiced to validate predisposed conclusions that support Chandra's premises and recommendations. Archarya was on her graduate committee, and in her thesis she thanks Chandra for "developing and making [the] project a reality."

The test water used for the student's experiments was from the Tahoe Keys Marina and wasn't representative of Tahoe's water quality or nutrient content. The water's calcium (13.6 ppm) was 50 percent greater than Tahoe's nearshore average, and instead of being maintained at or near Tahoe's highest monthly average of 64.7°F, the test water was kept at room temperature and pH rose significantly above Tahoe's near-neutral pH—the test conditions were constructed by Chandra and his associates and were unrepresentative of Lake Tahoe, but were elevated well above minimum levels for quagga survival (see Table A, pg. 56).

In its opening paragraph, the student's thesis states: "Today, quagga mussel specific research is still lacking and the physicochemical characteristics of aquatic systems required by quagga mussels to successfully establish is not fully understood. This includes an absence of research in aquatic environments in the western United States and on quagga veligers." Ironically, the thesis cites—and frequently contradicts, misstates, and omits more relevant data—from 50 scientific works, including 16 works devoted to western environments. Tellingly, it does not cite *Monitoring and Control of Macrofouling Mollusks in Fresh Water Systems,* the recognized authority on dreissenidae—a text that cites the contributions and work of more than 150 limnologists and marine biologists, but includes no one from Tahoe or UNR.

The most critical investigation in Chandra's $338,000 study for the BLM was assigned to a graduate assistant who performed her tests on the UNLV campus, which is 400 miles distant from Chandra's Reno lab. After conducting her research and tests, the student repudiated the accepted findings of the world's leading aquatic biologists, and her thesis concludes: "Data indicates there is the high possibility of quagga establishment in Lake Tahoe. It is recommended that boat inspections and [hot water] washings be continued to help prevent quagga introduction. Western lakes and streams should take the same precautions, despite having lower calcium concentrations [than Lake Tahoe]."

After the 2014 thesis was archived, when her Examination Committee Co-chair, Kumud Acharya, was asked when the report and data for the quagga study that his student had done the primary research on and that he and Chandra were co-authoring would be available, he said that a manuscript was "nearly complete and about to be submitted" and to check with Sudeep Chandra. Chandra didn't reply to a request for a date and source.

Chandra and Acharya's FY 2015 annual report to the Bureau of Land Management was due September 30, so that it could be posted by the end of the calendar year. As of June 2016, the final report and quagga risk assessment for Lake Tahoe, which had been twice previously postponed, was not posted on the Bureau of Land Management's website. Considering that 18 months earlier Chandra had released the five-year study's results to the *Reno Gazette-Journal,* and a year earlier Acharya had said it was nearly complete, it appears that the researchers don't want their work to come under public scrutiny.

Three months after Chandra announced through the *RGJ* that quagga could colonize in Lake Tahoe, TRPA's 5th Annual Tahoe AIS Forum was held in May 2015. Chandra not only didn't present the most significant findings of his career, he didn't attend the forum. Instead, he sent a short video in which he said that tests showed that quagga veligers were able to mature and settle in Lake Tahoe—remarkable findings that deserved to be presented to a more formal group than fewer than 40 Tahoe agency managers and interested parties. A request to TRPA for a download of the video was answered by Aquatic Resources Manager, Dennis Zabaglo. He said that the presentation had to be sent by Chandra. The UNR professor didn't reply to the request for a copy of his video, which has been posted nowhere.

The *Tahoe Daily Tribune* reported that at the forum, Nicole Cartwright, Tahoe RCD's AIS program coordinator, said invasive species are non-native species and "cause devastating effects to our economy and recreation," and that this resulted in "one of the most rigorous boat inspection programs in the United States." Zabaglo said that Asian clam eradication was "complex and expensive" and multiple methods would be tried in the future. The remainder of the article centered on non-native weed control in the Tahoe Keys and Emerald Bay—no mention was made of Chandra's presentation video or the remarkable claims from his study that quagga can colonize and propagate in Lake Tahoe.

Chandra was the opening speaker on the education day at the Western Regional Panel on Aquatic Nuisance Species' August

2015 annual meeting. The convention was hosted by TRPA and brought together 80 AIS program managers from the western states and Canadian providences. The gathering of knowledgeable biologists and aquatic science experts would have immediately seen the flaws in Chandra's study if he told them the same things he told the *RGJ*, so he changed the data and results of his unpublished study's parameters.

Even though the veligers in the survival tests were kept in water at room temperature, Chandra told the ANS Panel the test water was 66°F—significantly greater than Lake Tahoe's average summer temperature—but slightly above the veligers' minimum survival temperature of 64°F. Chandra avoided mentioning any of the study's other environmental controls, but during the question and answer period, an attendee asked what the lake's average calcium is. Chandra avoided a direct answer and replied: "People have told us, 'Oh, the average calcium at Lake Tahoe is 11.5, so there is no risk.' I think it is very dangerous to talk about averages. If you look at an average for Lake Tahoe it is 11.5 [a 25 percent exaggeration above the level he established]. If you go to the clam beds, it's 18 to 24 [a 180 to 240 percent exaggeration]." In the *Quagga Mussel Risk Assessment*, he reported that the highest calcium reading above a Tahoe clam bed, where veligers free-float for a month before settling to the bottom, was 10.0 ppm.

That question was followed by another that asked what Lake Tahoe's pH is, an important variable because it's been demonstrated that adult quagga can survive for up to 90 days in water with calcium as low as 11.5 ppm if pH is above 7.5. Chandra answered that Lake Tahoe's pH is between 6.8 and 7.2—he failed to tell the questioner that in his study's veliger tests the pH fluctuated between 7.5 and 8.06.

Steve Urie

— 5 —

Verification & Validation

Invasive species management requires prioritization, which should be based on informed and quantified assessment of the potential ecological and economic costs of species (both positive and negative), considered in the proper context of the invader and ecosystem.

~ Claire McLaughlan, Zoologist

A primary axiom of science is that until a theory is proven by real world physical evidence, it is not accepted as scientific law. In 1916, as a corollary of his general theory of relativity, Einstein predicted that when super-heavy celestial objects like black holes collide, gravitational waves ripple across the entire universe. Because black holes are (as their name implies) "black," they can't be seen by telescopes; consequently, even though astrophysics relies on the presence of black holes to explain the mechanical laws of the universe, for a century after Einstein theorized that gravitational waves from the collision of black holes would eternally ripple across the universe, there wasn't even proof that black holes existed.

On September 14, 2015, a team of researchers proved not only that black holes exist, but Einstein's theory that gravitational waves are created by their collisions became scientific law. In a decade-long experiment, the scientists recorded the sound of two black holes that violently merged 1.3 billion light years from Earth.

Because life on Earth is far easier to observe than colliding black holes, ecological conditions and processes that are initially demonstrated in the lab aren't accepted as factual until observed in nature. It's impossible to replicate natural habitats, climatic cycles, micro-environments, and food webs in a laboratory, which is why instead of relying on short-term experiments that test for isolated factors, research ecologists analyze years of field data from comparable ecosystems to verify their theories, lab results, and observations. An example of aquatic science that not only considered the entire body of relevant science, but validated its conclusions by comparing experimental results to field data is the Oregon State University study performed for the U.S. EPA. *A calcium-based invasion risk assessment for zebra and quagga mussels* evaluated 15 major dreissenid mussel studies and then compared consensus calcium requirements to data gathered over 5 to 10-year periods from 3,091 U.S. rivers, streams, and lakes.

In stark contrast to the OSU study are Sudeep Chandra's studies on potential quagga survival. No field data was analyzed, and he didn't cite any examples where quagga had colonized in an ecosystem where water is either as cold or as calcium deficient as Lake Tahoe's. The reason he didn't is that worldwide there is no lake ecologically similar to Lake Tahoe where dreissenidae have been detected, much less supported a sustaining colony. And unlike the Oregon State study, which used corroborating studies to verify established survivability and colonization thresholds, Chandra's conclusions relied solely on lab data collected by a graduate assistant and included no confirming or even similar studies—he stands remotely alone in his belief that quagga can survive long term in water with calcium below 12 ppm.

Biology examines life from the molecular level to human anatomy, and its complexity demands layers of specialization.

Biologists are divided into the areas of cellular biology, the examination of life's basic building blocks; physiology, the examination of the physical and chemical functions of tissues, organs, and organ systems; evolutionary biology, the examination of the processes that produce life's diversity; and ecology, the examination of how organisms interact in their environment.

At the end of one of ecology's branches is limnology, the study of freshwater life science. Closely related to aquatic ecology, limnology studies the relationship of aquatic organisms. It covers the physical, biological, and chemical attributes of lakes, rivers, streams, and wetlands and is the science that forms the basis for the conservation of freshwater ecosystems. The field is diverse, and scientists are only beginning to understand the complexities of ecological interdependencies and interactions.

Peer review is at the heart of the process of all science. It is the method by which grants are allocated, papers published, and academics promoted. Ideally, peer-review incorporates the evaluation of data and studies by multiple independent and similarly competent reviewers. And so that undue influence is avoided and impartiality is maintained, the use of colleagues from the same university or institution for review is discouraged. The first questions editors ask journalists and science writers when they quote a study is: "Was it peer-reviewed, and by whom?"

Since the 17th century, the process has provided a form of scientific self-regulation that helps assure standards of quality are maintained and results and conclusions are credible. Although not foolproof, when university studies that are funded by public grants are peer-reviewed, only one-half of one percent of the conclusions are later retracted. However, when published in a "scholarly" journal (a periodical or website that archives and prints studies by experts) the percentage of findings that are ultimately proven wrong soars to almost a third. Not only are none of Sudeep Chandra's AIS studies peer-reviewed, and all disregard accepted science and repudiate studies that were.

Chandra received a $338,000 grant to determine whether quagga mussels could propagate in water with less than 11.5 ppm of calcium. No study or environmental data has shown that a sustaining quagga population can develop in water with less than 15 ppm and pH under 7.5—165 percent greater than Lake Tahoe. If he had demonstrated that quagga could survive long-term in water with calcium less than 11.5 ppm, it would have been a singular accomplishment, but he didn't even test to see if either adult or veliger quagga could survive in water with calcium less than 13.5 ppm. Whenever unprecedented lab results are obtained that are outside of accepted science, it is common procedure to redo the experiment at least once to verify the findings. To not do so disqualifies the study from respected peer-viewed journals.

Scientists from TERC, DRI, and UNR created their own journal and repository for Tahoe's scholarly papers, symposiums, and publications. Beginning in 2006, the Southern Nevada Public Land Management Act provided $2,703,000 to create the Tahoe Science Consortium. Headed by TERC and apparently unclear on how the peer review process works, TSC states that its "purpose is to provide peer review, advance ongoing science efforts, develop a research plan, and other scientific activities to assure quality science and projects that serve as a basis for decision-making in the Tahoe Basin." The first papers archived by the virtual institute were in 2007. A June 2016 visit to TSC's archives revealed that no scholarly work had been cataloged since 2011.

A fifth-year UNR professor when the quagga mussel scare gripped Tahoe's environmental community in 2007, Chandra was one of two university-affiliated limnologists within 100 miles of Tahoe. He was a logical choice for TRPA to contact for advice on how serious the mussel threat was and how to best protect the lake, and he became a willing media contact. Within months of quagga invading Lake Mead, the glib professor opportunistically seized the lead in analyzing the potential threats to Tahoe by all non-native species and forming the protective response to them.

Scientific fraud covers a wide spectrum of professional misconduct, ranging from serious offenses such as fabrication, and falsification to lesser misconducts such as deceptively interpreting data or exaggerating or misinterpreting results to provide unjustified conclusions that serve professional, financial, or political purposes. In *The Great Betrayal: Fraud in Science,* Nobel Laureate Horace Freeland Judson writes: "Fabrication is faking data entirely; biologists call it 'dry labbing.' Falsification is manipulating data that have been obtained—picking only the favorable results, trimming off data points that threaten the paper's conclusions, presenting readings barely above background levels as significant, combining the best of two experiments into what then appears to be one, and so on through all the permutations the imp of the perverse can whisper."

Peer review usually prevents scientific fraud, and without it, deviations from accepted practices in proposing, carrying out, or reporting results can go undetected. Chandra doesn't fabricate data, but unquestionably, he falsifies data by manipulating data collection, misstating data points and background levels, and staging experiments to achieve desired outcomes. And he uses his and his colleagues' academic positions and credentials to lend credibility to unwarranted public policy recommendations.

Scientific studies that return exceptional or antithetical results that later prove to be wrong are often dismissed as simply being uninformed or sloppy work. Because their lab work seldom makes it into the scientific sphere and is monitored by their instructors, erroneous data or research results obtained by a student is acceptable, but when a research professor manipulates student-work to prejudice data for publicly-funded grants and studies that justify multi-million dollar programs, it is scientific fraud—and this is what Chandra has done.

Lake Tahoe's most influential researcher greatly distorted background levels by supplying non-representative test water to a student and set unrealistic and unacceptably broad controls and parameters so that she unwittingly produced fraudulent data and drew irresponsible conclusions. He then used the corrupted data as the basis for his groundbreaking announcement that quagga mussels could not only survive in Lake Tahoe-quality water but also that their veligers quickly grew into reproductive adults. Although probably meant gratuitously, the student aptly characterized Chandra's survivability study when she thanked him for developing her thesis and named him as a "partner in crime."

After releasing the *Quagga Mussel Risk Assessment*, Chandra said, "Using sound science and good policy, [Lake Tahoe] can be protected." But his science wasn't sound or even original. He duplicated a 15-year-old study and got the same results, but he then slanted his conclusions to justify already implemented policy. And he created uncertainty where none had existed. If an ecosystem can't support quagga mussels, decontaminating boats that potentially transport them is foolishly unnecessary.

The *Quagga Mussel Risk Assessment* greatly exaggerates the economic damage quagga do in southern Nevada reservoirs. The experiment's report quotes Peggy Roefer, Regional Water Quality Manager for the Southern Nevada Water Authority — Las Vegas' water supplier—as saying the utility "spent approximately $32 million to manage quagga mussel impacts on the water intake infrastructure."

SNWA has two 20-foot diameter water intakes, and because of Lake Mead's rapidly falling water level, a third intake is being constructed 200 feet below high-water. The first contract for the $489 million project was for a 500-foot tunnel that was completed in 2010. It cost $32 million. In 2008-2009, SNWA did spend $32 million for water intake infrastructure, but it wasn't to mitigate quagga mussel damage.

Roefer explains that after divers found a room-size pile of quagga shells that was 2-feet deep, the shells were removed. She reported that the utility found they were able to treat raw water from Lake Mead with chloramines, which even in highly diluted concentrations are lethal to quagga larvae. The treatment method kills veligers before they can develop and attach to the intake pipes, and along with other impurities, the larvae are removed from the water during the filtration process.

Roefer couldn't explain how Chandra came up with his $32 million figure and said, "Very little damage is caused by the quagga mussels. At our facility, the annual cost for treatment and shell removal is about $150,000." The SNWA processes some of the nation's warmest, most calcium-rich water and has the greatest dreissenid mussel problem of any U.S. water utility.

Because calcium concentration affects the density of dreissenid colonies and water temperature affects reproduction cycles, control measures that are effective at one facility don't necessarily transfer to another. In southern Nevada reservoirs, quagga reproduce two or three times a year, whereas in Lake Erie, where calcium is lower and water is colder, they reproduce once or twice a year. And Roefer says various methods and levels of control work in different water treatment plants. The methods include chlorination, commercial molluscicides, injection of liquid carbon dioxide, lowering pH, treatment with copper sulfate, and ultraviolet light, a method used at Hoover Dam (CSGE, 2012).

By the summer of 2009, environmental agencies were reeling from the Great Recession. The U.S. Forest Service's budget for FY 2009 was slashed by 22 percent, but they found the money to almost double their contribution to Tahoe RCD's Invasive Species and Biological Resources Program. TRPA contracts Tahoe RCD to operate the Tahoe boat inspection program. The U.S. Bureau of Reclamation's Water and Related Resources budget was reduced by 11 percent, but they were able to increase their contribution to Tahoe RCD by 28 percent. And Nevada, the hardest hit of all states by the depressed economy, found $6,311 for the Tahoe RCD inspection program after giving nothing in FY 2008—for FE 2015 and 2016, Nevadans are matching Californians and annually contributing $375,000 for Tahoe boat inspections.

Funded by money obtained from federal land sales in Southern Nevada, TRPA was unaffected by environmental belt-tightening during President George W. Bush's administration and the 2008 Great Recession. Well financed and with scientific backing from a series of highly questionable studies, the federal agency launched a relentless public information campaign in support of their war on invasive species.

Quagga mussels became the primary target for Tahoe's AIS awareness and prevention programs. Before the 2009 boating season, press releases were sent to area news outlets that read: "Boat inspections were implemented last summer to prevent quagga mussels from becoming established in Lake Tahoe. The mussels have spread to other bodies of water in Nevada and California. Once established, scientists say there is no way to get rid of them, and they seriously alter the aquatic ecosystem.

"Lake Tahoe environmental regulators are considering a number of options, including boat inspection fees, to fund efforts to keep invasive mussels out of the scenic lake. ... 'We're trying to protect the lake, and in order to have an inspection program, we need to have a consistent, long-term funding source,' said Ted Thayer, TRPA's science team leader."

After the 2009 summer boating season, Thayer said, "Even though winter's approach means there are fewer boats on Lake Tahoe, any watercraft launched will continue to be inspected for the presence of invasive mussels. Quagga and zebra mussels pose a big environmental and economic risk to the lake, and vigilance is key because the mussels can spread just as easily in the winter as in the summer."

Thayer's message was incorrect. Water temperature is a critical factor during quagga development. Quagga veligers don't mature if water temperature doesn't average at least 64°F for the month they develop into juveniles. (Mackie & Claudi 2009) And only in some of Lake Tahoe's shallow waters do summer surface water temperatures briefly rise to that average.

Speaking about the necessity of winter inspections, Thayer said, "It's a big deal and certainly a high priority. The threat to the regional economy is considerable, and it is not an imagined threat." But it is an imagined threat. Disputed by all research except Sudeep Chandra's, TRPA escalated their war on aquatic nonnatives, and instead of a threat to the economy, it appeared that generous funding for unnecessary research and AIS prevention and control programs was created at taxpayer and boaters' expense.

Overseen by TRPA and implemented by Tahoe RCD, during the 2009 boating season 14,000 boats were inspected before launching, 600 were deemed to be not clean, drained, or dry and were decontaminated with bleach. Projected to cost about $650,000, the inspection program instead cost $1.2 million. Four years later, the number of inspections remained about the same, but decontaminations jumped from 600 to 4,221, and a $35 hot-water decontamination fee was added.

From the release of the *Quagga Mussel Risk Assessment* and Chandra's claims that Asian clams create calcium "hot spots" and facilitate quagga, harm biodiversity, and cause algal blooms until their own studies showed those claims to be false and the their Asian clam control program crashed, TRPA steadfastly supported their primary AIS researcher and defended his science. But when criticism of Chandra's science that supported their AIS policy grew louder, public demands for an independent AIS risk assessment were first heard.

It appeared that TRPA Board Chairman Casey Beyer responded to the calls for a re-evaluation of TRPA AIS policy when he wrote in a March 2015 op-ed in the *Reno Gazette-Journal*: "A major priority for TRPA this year is completing a comprehensive control strategy for aquatic invasive species at Lake Tahoe. The goal is to prioritize control projects and guide our future efforts to best use limited funding for maximum possible effect." The possibility of an objective re-evaluation of Tahoe's AIS strategies brought hope to the AIS program's critics.

The promise for a new AIS strategy crashed four months later when the *Implementation Plan for the Control of Aquatic Invasive Species within Lake Tahoe* was released. The plan was written by Marion Wittmann and Sudeep Chandra. Seven agency managers and academics were appointed as an "expert review panel" to evaluate it. Disappointingly, the panel had little of substance to review. The plan didn't evaluate whether there are any plants or animals that may be an invasive risk at Tahoe, and New Zealand mudsnails and spiny water fleas weren't evaluated and aren't even mentioned—nor are quagga mussels.

No evaluation is made of the boat inspection program, and the plan only addresses invasive species prevention by saying: "There are a number of AIS prevention efforts currently in place in the Tahoe Basin. The most significant prevention program is the boat inspection and decontamination program. Began in 2008, the program is managed by TRPA and its designee, Tahoe RCD. The objective is to completely eliminate all viable AIS life stages, and thereby prevent their introduction into [Tahoe] waters."

The implementation plan addresses seven established species that Wittmann and Chandra claim are invasive to Tahoe and states that there are no viable control options for two of them: Asian clams and mysid shrimp. Two others potentially may be controlled: bullfrogs and crayfish; but the plan recommends no immediate action for them. The only "invasives" that the plan targets for control are Eurasian watermilfoil, curlyleaf pondweed, and warm-water fish. Most of the report is devoted to justifying and prioritizing the control of the two plants and warm-water fish in the Tahoe Keys.

The plan reveals no information or science that wasn't previously known, and issues that were hoped to be addressed weren't. Currently, no plant or animal on the USDA's list of aquatic invasives has been shown to be able to propagate in Lake Tahoe and justifies having boat inspections. And outside of the Tahoe Keys, there is no plant or animal that causes harm and needs to be controlled. But by announcing the study weeks after the Asian clam control program folded, TRPA bought time to allow the disastrous project to fade from the public consciousness and to deflect new calls for an independent AIS risk assessment.

Even though the Implementation Plan emphasizes that there is "no feasible option" available to control Asian clams, three months after the plan was released, in October 2015 TRPA Aquatic Resources Program Manager Dennis Zabaglo was asked if there were any future plans for the control of Asian clams. He replied: "Next month, we are working with a company to test their new technology for Asian clam removal."

Steve Urie

— 6 —

An Ounce of Prevention

Without the boat inspection program, Tahoe would be at risk to all sorts of invasive species. A lot of things can happen to negatively impact our ecosystem, our environment, and our local economy as well.

~ Ted Thayer, TRPA AIS Program Coordinator

On first read, Ted Thayer's statement seems plausible, but boat inspections wouldn't have prevented the introduction of any of Tahoe's 28 aquatic nonnatives. And the few that have escaped from the Tahoe Keys and thinly populate spotty habitats along the nearshore where sufficient food and cover is found haven't had a negative impact on the lake's ecosystems. Most Tahoe anglers believe that introduced game fish and the crustaceans they feed on have improved the lake's environment and have unarguably helped the economy. But more importantly, no aquatic nonnative found anywhere in North America has been identified that can propagate in Lake Tahoe and measurably harm it.

Although inaccurate, Thayer's message was much less inflammatory than TRPA's media blitz that preceded the first mandatory inspections. An early press release said: "The words 'aquatic invasive species' should strike fear into the hearts of people who love Tahoe's pristine waters—especially the words 'quagga' and 'zebra mussels.' If even one larva of these invasive mollusks enters the lake, it could multiply rapidly and the resulting population could destroy Tahoe's environment and economy. These water locusts can attach to almost any hard surface and can clog water infrastructure and boats. Once in a body of water they eat through the food column; one adult mussel can filter through a liter of water a day and remove phytoplankton[7] and zooplankton that is necessary to support Tahoe's other species.

"They have never been removed successfully from a large body of water. Given these consequences it's hard to understate how important it is to comply with TRPA's boat inspections this summer. Inspections are part of life if we want to keep the lake open to boating and recreation." The choice of closing the lake to boating or accepting inspections was not lost on boaters, and inspections seemed to be an environmental responsibility. But the premise was wrong—one quagga or zebra mussel larva (or even billions of their larvae) would die within days in Lake Tahoe.

The assertion that boat inspections protect a lake from quagga was dealt a setback in early 2013 when the mussels were found in Lake Piru, a 4-mile-long reservoir north of Los Angeles—a lake that had an aggressive inspection program. Lake Piru was the first California lake or waterway, not fed by the Colorado River, to be infested. It won't be the last. Mussels were also detected below the Piru Reservoir dam in Lower Piru Creek. It's inevitable that mussel veligers will be carried by currents and various other forms of transport the length of the Santa Clara River, and within a decade or so, quagga will establish in all California water bodies with favorable habitat. Ironically, boaters are now asked to make sure their craft is clean, drained, and dry after leaving Lake Piru.

---

[7] Microscopic aquatic plants

In November 2007, Casitas Municipal Water District Board President Russ Baggerly said quagga are "a serious threat. If just one gets into a water system, it could multiply by the hundreds of thousands and cripple infrastructure, decimate fish populations, and degrade water quality." Baggerly's concerns were shared by most water managers west of the Rockies. But because calcium in southern California's freshwater systems range from moderate to high for quagga infestation, there was solid basis for SoCal water managers' fears. However, as in Nevada and the Great Lakes area, Baggerly's worries about crippled infrastructure and decimated fish populations didn't happen—and won't.

It took mussels less than five years to colonize susceptible Colorado River System habitats. The Colorado River Aqueduct diverts water 240 miles to Metropolitan Southern California, and water from the 80-mile-long All-American Canal irrigates the Imperial Valley. Quagga are in each of the waterways. After the mussels were detected in Lake Mead, mandatory inspections were instituted in most major southern California lakes—within seven years quagga were detected in 33 lakes and reservoirs.

Southern California authorities had believed that boat inspections would prevent dreissenidae from spreading from the Colorado River to uninfested waters, and for a few years they appeared to work. If boats were the only way AIS jumped from lake to lake, decontaminations before launching might be an effective preventive measure, but they aren't. Microscopic eggs and larvae are carried by stream flow and are spread by migratory birds and animals, personal watercraft, anglers' gear and tackle, and even wet swimwear.

A remarkable example of how animals spread worldwide occurred in April 2014 when Asian mussels were discovered on a British Columbia beach clinging to a piece of tsunami debris from Japan's 2011 earthquake. Certainly, cleaning, draining, and drying boats has slowed the rate of spread of AIS, but as dreissenidae and other non-native aquatic species have relentlessly swept across the Northern Hemisphere, there is no evidence that boat inspections and decontaminations have stopped any non-native infestation anywhere—the only thing that prevents invasions is when nature deprives the non-native species of a suitable habitat.

It is not surprising that most kayakers and paddle boarders aren't familiar with the biology of dreissenid mussels, spiny water fleas, and New Zealand mud snails and that they don't question what they are told by TRPA. But it is surprising—and disappointing—that those charged with protecting Tahoe's aquatic habitats and water quality would ignore the process of biological investigation and analysis. Worse, is that the seldom fact-challenged academics who advise the agencies would distort their data and conclusions and make recommendations to satisfy the expectations of those who commissioned them.

Chandra and TRPA charted an intractable course when at the height of the 2009 summer tourist season they announced the results of the *Quagga Mussel Risk Assessment*. A TRPA press release stated: "Scientists say a new study shows invasive quagga mussels can survive and possibly reproduce in Lake Tahoe. 'This could potentially be catastrophic for the lake,' said Ted Thayer, natural resource and science team leader for the TRPA.

"Chandra [said,] 'It's going to take a little more research to determine if this mussel can really sustain in populations over time.' If established at Tahoe, quagga mussels could cause profound changes to the lake's sensitive ecosystem. The mussels could clog water intakes, cover boats and piers, and litter pristine beaches with sharp and reeking shells. A recent study by the U.S. Army Corps of Engineers estimates a mussel invasion could cost Tahoe's tourism economy more than $22 million per year.

"'This is disappointing,' Thayer said. 'It would have been nice if it had come back that there wasn't that high a risk.' TRPA and other agencies have taken steps over the past two years to prevent introduction of the mussels, including inspection of all boats launching into the lake. Chandra said he believes boat inspections can be effective. 'When you reduce the likelihood of transfer, almost always those ecosystems are protected.'"

Five years later, hedging his position that inspections are effective Chandra said, "Inspections are only as good as the people driving them. Some species, such as plankton and mussel larvae veligers, are microscopic. You can't see them, so filtration is a better way to keep them out [of the lake]."

Chandra's remarks were made to promote the Mussel Mast'R, a device he designed in collaboration with Wake Worx. Apparently no longer agreeing with his prior premise that "If even one larva of these invasive mollusks enters the lake, it could multiply rapidly and the resulting population could destroy Tahoe's environment and economy," Chandra said, "By using the filtration system, we can prevent 97 to 99 percent of small species such as the quagga mussel out of the boat's ballast water using this filtration system."

After serving as the agency's legal counsel for two years, in February 2009 Joanne Marchetta was appointed TRPA's executive director. At the start of the boating season, in May she wrote: "This is the season boaters and water recreationists wait for. But appearances can be deceiving. Over a number of years, invaders have entered the lake: Species that are not native to Tahoe have affected the lake's ecosystem. ... Quagga and zebra mussels are carried between water bodies by boaters. Currently, there is no threat more imminent to the lake than the risk of infestation by these aquatic invasive mussels.

"The stakes are high, and failure is not an option. To fail means unimaginable environmental and economic consequences. This summer we are increasing our vigilance and are asking even more of Lake Tahoe boaters. All launch ramps will be staffed with boat inspectors trained to search for invaders like the quagga and zebra mussels where they hide."

The same Zephyr Cove resident who said we are losing the war on dreissenid mussels wrote: "Inspections won't work. They won't protect Tahoe from the zebra/quagga. Over the last decade, TRPA has spent $1 billion on environmental quality improvement in the Basin, but now we are told they are forced to charge fees to pay for boat inspections. … We must stop wasting time and money on locals and focus on the real threat: boats coming from zebra-infested water."

Another resident differed with that viewpoint: "There are new rules for Tahoe this summer due to the possible invasion of quagga and zebra mussels. … The inspections are laughable at best. … What about Donner, Boca, Stampede? Apparently these bodies of water have not been deemed by any governing agency important enough to protect. There are no inspections on any of these other lakes—they are on their own and at risk."

Although at Tahoe probably thousands, if not millions, of quagga larvae had been introduced into the lake but soon died in the years before boat inspections began, the new program was bolstered by Chandra's *Quagga Mussel Risk Assessment*. Ted Thayer said: "With a recent study showing quagga mussels could survive and possibly reproduce in Lake Tahoe, [we] are confident current inspections are adequate to keep out the invasive mollusks. We've been working very closely with researchers before they were able to put out a final draft so we were aware of what was going on. What we are doing right now is sufficient in light of this new research."

Nicole Cartwright, head of the Tahoe RCD Invasive Species Program, said: "When we wrote our procedures and protocol we weren't waiting for this info. Every action we take is with the knowledge that [quagga] could survive in Tahoe." Considering all previous data had shown that quagga couldn't survive in low-calcium water and that even Chandra said he was surprised by his experiment's results, it was more surprising that TRPA and Tahoe RCD had known all along that quagga and zebra mussels could survive in Lake Tahoe.

Two months later, the boat inspection program received a huge boost when Congress earmarked $2 million of a $32.2 billion environmental protection bill for Tahoe boat inspections. TRPA said about $800,000 would be used to study how AIS inspection stations could be moved to roadways instead of at lakeside launch sites, with the remainder used to fund operations. Even though the inspection program was justified by a study that cost $1/40^{th}$ of what was going to be used to study the relocation of inspection stations, and TRPA still said, "It remains unclear whether Tahoe's physical habitat can support [quagga mussel] establishment," none of the $2 million was used for research to determine if the inspection stations were even necessary.

Four years after saying more research was needed, in early 2013 Chandra reaffirmed that he had indeed began that research. That May he told the Town of Truckee, "I am looking forward to sharing with you our final results from these new quagga survival experiments we are conducting with Tahoe water later in the year." When the year passed without receiving results from the follow-up study, he was asked the status of his research.

Chandra wrote: "Regarding the experiments, they have not been concluded due to delays in the project, but we are starting our second set of experiments in March [2014]." He didn't say why delays would require a second set of tests, much less what the results of the first set were, but he later reported to the U.S. Forest Service that his "experiment [was] complicated by the abrupt die-off of mussels."

He also wrote: "Concerning invasion ecology [and its] uncertainty, as I have emphasized before, focusing on mussels or their potential for colonization should not be the only reason to establish [Truckee's boat inspection] program. There are other taxa that may pose a threat to the region. Prevention is better than the uncertainty of invasion. ... We are presenting our results to date together with the uncertainty before the completion of the final experiment." In the following year, Chandra didn't present his results, or was it shown then or since then that any non-native aquatic taxa that is capable of establishing in Sierra lakes and streams can cause environmental or economic harm.

In July 2014, five years after she had written that "there is no threat more imminent to the lake than the risk of infestation by [quagga] mussels," Joanne Marchetta repeated her dire warnings in a regional op-ed. She wrote: "The stakes are high. To fail could mean unimaginable environmental and economic consequences." She reported that, "We've kept new destructive invaders out of Lake Tahoe over the last five years—the program has resulted in 36,000 inspections, 14,000 decontaminations, and unprecedented public awareness of the threat of invasive species." It was observed by some that during the same period there were no new invasions in any of the more than 150 large Sierra Nevada lakes that have boating access and don't have inspections, and that the unprecedented public awareness was based on false science and TRPA's disinformation.

Marchetta also reported that "TRPA is leading a public-private partnership ... to secure funds to keep the watercraft inspection program going next year. Boater fees cover half of the

$1.5 million cost of the annual invasive species prevention program and federal funds historically covered the balance. This federal funding is drying up and the program is in urgent need of new revenue sources to continue inspections."

Two months later, after securing funding for the 2015 boating season from another agency, Marchetta wrote, "Tahoe is standing on a fiscal cliff and the ground is sloughing off beneath our feet. Finding ways to secure funding is our biggest challenge." For less than what boaters annually pay in inspection fees, TRPA could perform an in-depth AIS risk assessment, and would almost certainly discover that they could save more than $4 million a year by scrapping their aquatic invasive species programs.

Four million dollars a year is almost inconsequential for TRPA. In July 2015, after being rejected by Congress in 2009, 2011, and 2014, the agency again sought $415 million over the next decade for many environmentally important efforts, including watershed and creek restoration, forest management to reduce the risk of wildfires, and treatment of wildfire burn areas.

Of the $41.5 million a year TRPA has requested from Congress, $4.5 million is earmarked for AIS prevention and control. The agency's annual cost for the boat inspection program is $750,000 "to prevent introduction of quagga mussels," leaving $3.75 million a year to control "curlyleaf pondweed, Eurasian water milfoil, and warm-water fish like bluegill and bass"—the only aquatic nonnatives that Wittmann and Chandra recommend controlling—and which more than 90 percent are contained in the Tahoe Keys.

Steve Urie

— 7 —

A Pound of Cure

One of the disadvantages of [impervious bottom barriers] is that you see death just about everywhere. It takes out the good guys along with the bad guys.

~ Geoffrey Schladow, TERC Director

In 2008, while Sudeep Chandra was theorizing how Asian clam shells could raise calcium levels in Lake Tahoe and enable a quagga mussel infestation, UC Davis researchers were testing methods to control the clams. A $400,000 experiment to test two ideas how to eradicate the clams and contain them at the South Shore was announced in a December press release: "While Asian clams cannot attach to as many surfaces [as quagga mussels], they can travel in water currents during their larval stages. According to research, currents in South Lake Tahoe could take the clams to the West Shore, Marion Wittmann said.

"In January, research teams will begin testing two non-chemical [methods to] eradicate the clams. Because the clams are sensitive to oxygen content, one method will try to suffocate the clams by placing a tarp over the beds. The other method will be to vacuum the clams. Wittmann said the good news about the clams is that they haven't completely invaded the lake. 'The system isn't saturated with clams yet, hope is not lost.'"

Of the two methods tested, smothering the clams was less environmentally destructive and more cost efficient than sucking them up and picking them out of the sediment. After tracking the clams' survivability, Wittmann reported that after 28 days at peak summer temperatures, 100 percent of the clams were dead, and that all life under the mats, including a dozen species of native plants and animals, were killed in the test. Wittmann said, "That's a potential problem," and rhetorically asked, "If it has an impact on the entire ground-dwelling community, is it worth it?" More than a century ago, John Muir had answered her question when he observed: "When we try to pick out anything in nature by itself, we find it hitched to everything else in the universe."

Some of the detrimental effects from scouring the bottom of Lake Tahoe of most life are well known, and they are extensive. Chris Rosamond, a freshwater biologist at Nevada's Desert Research Institute (DRI) wrote in a *Moonshine Ink* op-ed: "I concur [with the] disdain for the 'Keep Tahoe Blue by putting big blue tarps all over the bottom and killing everything' approach that has been used these last few years.

"In addition to the initial shock of suffocating whatever native benthic organisms there are in these locations, the anoxic conditions under these tarps will alter nutrient dynamics in the benthic substrates, particularly phosphorus and nitrogen cycling. Recall that nitrogen and phosphorus are our two greatest pollution concerns in the Basin. In short, sterilizing the floor of Tahoe's shallow habitats through suffocation is probably not a viable long-term solution." (Rosamond, 2013)

Rosamond didn't realize the prescience in his op-ed. Two years later, TERC reported that research revealed that while Lake Tahoe's clarity is controlled by fine particulates, blueness is most affected by algae, and when algal growth decreases, Lake Tahoe appears bluer—phosphorous and nitrogen fuel algal growth.

Asian clams burrow into bottom sediment, where they lead uncomplicated lives lasting from two to four years. They are asexual and require temperatures above 63°F for larval release. Tahoe's average surface water temperature is less than 52°F and only in some shallow waters during the summer do temperatures reach the mid-60's—a condition that explains why Asian clams don't populate Tahoe nearshore where temperatures are colder than in South Shore shallows or colonize to the same degree as they do in warmer lakes. Water temperature, nutrients, pH, calcium concentration, and substrate play integral parts in determining how successfully Asian clams colonize. Proof of how hard it is for them to prosper in Lake Tahoe is evidenced by their size—there they grow to about half the breadth and a quarter of the weight of those in warm-water lakes that are rich in nutrients and have a hospitable, grainy lake bottom.

When the size of areas covered by the tarps increased in the control tests, their effectiveness decreased. *Tahoe: State of the Lake Report 2012* reported the results of a $648,000 follow-up pilot project on two half-acre sites at Lakeside and Marla Bay: "Synthetic rubber barriers were placed over the Asian clam infestations in the southeast portion of the lake to deprive the clams of oxygen. The barriers were removed after 120 days. Clams under the barrier were reduced by more than 98%." Because adults can self-fertilize and release up to 2,000 veligers per day, a 2 percent survival rate ensures that populations will sustain.

At the prior year's Lake Tahoe Aquatic Invasive Species Forum, the project's leader, Geoffrey Schladow, explained that Asian clams live under lake-bottom sediment, and seeking oxygen they rise from the sediment to the lake floor. And after dying, instead of being buried beneath as much as six inches of sand and gravel, their shells lie on the lake floor. He said pictures of shells on Tahoe's lake bottom were from clams killed by TERC projects.

Since 2008, a greatly enlarged underwater photo of one of the pictures that Schladow referenced has become ubiquitous. An image of algae-covered Asian clam shells on a sandy Lake Tahoe beach bottom has an apparently huge glob of algae floating above dollar-size shells—in fact, the mass of algae is less than a foot long. The distorted image of clam shells and filamentous algae, which was distributed as free computer-screen wallpaper in an app promotion, is explained by Schladow's description of how the clams rise to the lake bottom under the tarps seeking oxygen and Rosamond's explanation of how the anoxic conditions under the mats creates algae-producing phosphorous and nitrogen.

More than a dozen native aquatic plants and animals live on the floor of Tahoe's nearshore. One of the native species is the peaclam. About their chances of surviving slow suffocation under rubber mats, at the 2011 AIS Forum Schladow said, "Some of the good guys like the [peaclam] will possibly survive." Two years later, TRPA took Schladow's nonchalance towards Tahoe's native species a step further, and said, "Control actually gives the natives a leg up in the competition, even after killing some of them." As insensitive as TRPA's message was, it is accurate: Because they aren't as biologically adapted to the habitat, Asian clams recolonize more slowly than most native species. But it's not a competition. As Cornell researchers explained in their criticism of Tahoe's Asian clam management plan, the clams co-occur with abundant populations of Tahoe's native mollusks.

At the AIS Forum, Schladow reported the results of the two half-acre tests. He emphasized his lecture with PowerPoint slides. One showed that the rubber tarp method cost "$2.50 per sq/ft," and "100 clams were killed for each penny spent." Even though the test project cost six times his $2.50 per sq/ft claim, and there would had to have been 25,000 clams per sq/ft (40 times more than the reported infestation) to meet his projection, the attendees accepted his numbers and left the forum with the comforting feeling that rubber bottom barriers were an efficient, inexpensive method to control Lake Tahoe's Asian clams.

In October 2012, the *Sierra Sun* reported: "After the California State Parks staff discovered Asian clam shells on an Emerald Bay beach in May [2009], UC Davis conducted an underwater survey to document the infestation. Three years later, the Lake Tahoe Invasive Species Program, a coalition of 40 public and private organizations [began] an attempt to smother five acres—approximately 6 million to 12 million individual clams [25 to 50 per sq/ft; surveys later showed the density to be 5 to 6 per sq/ft]—of infested lake bottom near the mouth of Emerald Bay." Under Schladow's direction, TERC managed the $810,000 project.

Notable from the news report was that in 18 months, the cost of clam removal had skyrocketed. No longer were clams being killed at the rate of 100 per penny, but the cost had escalated to a dime each—1,000 times more than Schladow projected at the forum. That wasn't the first time the professor's numbers didn't add up. When the barriers were removed after 120 days at the pilot sites, the clams were reduced by 98 percent, and a year later their density was 90 percent of the original population. That seems impressive, until put in the context that in a year the clam population quadrupled, and at that growth rate would be back at 2010 levels in 2013—later population surveys revealed that it didn't take that long.

The AIS management plan that Schladow coauthored, says Asian clams "produce at least two cohorts per season." However, that was but a tiny fraction of what his colleague, John Reuter had claimed three years earlier when the clams' prolific fertility was used to justify TRPA's control program. Reuter told *UPI Science News*: "While the clams are small, they are fecund, producing up to 100,000 new clams during their four-year life span."

When asked about TERC's contradictory growth rates, Schladow wrote: "Regarding the growth rate, I believe that a more realistic rate of increase is the following year a further 8% may be affected, i.e. from 98% clear to 92% [sic] clear to 84% clear." He didn't reply when asked why he straight-lined the growth rate after telling the media that it was exponential, that test site results had confirmed that it was exponential, and that his assistant director had said that it was thousands of times greater than he projected.

Eight months after Schladow said that clam regrowth after two years would be about 16 percent, *Natural and Human Limitations to Asian Clam Distribution and Recolonizations—Factors that Impact the Management and Control in Lake Tahoe* was released. The $264,000 Forest Service study lists Wittmann, Schladow, Chandra, and Reuter as the primary investigators. The report says that clam regrowth after two years was nearly 100 percent—not 16 percent, as Schladow had predicted—and added: "Costs of Asian clam treatment in Lake Tahoe were approximately $210,000 per acre. Treatment [for 100 acres] can range [up to] $26 million and will neither achieve eradication, nor maintain low density populations of Asian clams in Lake Tahoe."

Simply, suffocating clams with rubber mats is ineffective and prohibitively expensive. The curiously named report was the concluding Asian clam study at Tahoe and reported the results from the researchers' series of tests and studies leading up to the Emerald Bay Project. The 78-page report addresses only Lake Tahoe's Asian clam management plan, yet 128 mostly inapplicable or irrelevant sources are cited. A third of the report is devoted to complex and pointless hydraulic flow analyses to explain how Asian clams "transport between shallow and deep [water] zones;" whereas, critically important data, such as repopulation results from each of the pilot projects isn't included or is incomplete.

In July 2010, *The Tahoe Daily Tribune* reported: "A bevy of local and national media outlets witnessed the implementation of the [Marla Bay] experiment. ... TERC Director, Geoffrey Schladow, rolled out a series of 100-foot by 10-foot black bottom barriers and spread them over a half-acre portion of the bottom of Lake Tahoe. Schladow said, 'The goal of this experiment is to determine whether it is feasible to control clams using impermeable bottom barriers. ... When it is all done, we must know whether the clams recolonize the treated areas.'" Two years later, he had his answer: surveys showed that populations in the treatment plots were nearly the same as in the untreated control plots.

Only weeks after the September 2012 population surveys were completed, Schladow directed the startup of the Emerald Bay project and predicted that the clams would repopulate by about 16 percent after two years. His report to the Forest Service had glossed-over the ineffectiveness of suffocating Asian clams under rubber mats and only hinted at the results recently learned from the repopulation surveys and ambiguously said: "Bottom barriers provide a promising method for reducing Asian clam populations, but fundamental unknowns present obstacles to recommending the large scale adoption of this methodology."

At the start of the project, Schladow said the goal was to control an "exploding population" of Asian clams [6 clams per sq/ft is inconsequential], and that "complete eradication just isn't feasible or even possible. You're laying down mats—there are going to be gaps between them. But control is possible, so the shells aren't affecting people and the clarity isn't getting worse. From a scientific point of view, I consider [the project] a success already. From the point of view of the lake, it'll be a success when there's a sustainable control program in place."

Schladow's research in the two years prior to the Emerald Bay project had clearly indicated that Asian clam control wasn't practical or economically viable, and most believe that a program isn't scientifically successful if it pollutes pristine water, kills native species, and does more harm than good. While watching a Tahoe RCD video of divers working in crystalline water and unrolling sheets of black tarps across Emerald Bay's sandy bottom, a single question comes to mind: "What are they thinking?"

Asian clams may have been in Lake Tahoe for more than a decade before they were noticed. They fill a minute ecological niche, and until justifications were needed for the boat inspection program, they were a non-issue. When asked about the potential extent of clam colonies, TRPA said, "While Asian clam infestations currently cover approximately 100 acres of Lake Tahoe, suitable habitat for Asian clam includes approximately 12,000 acres of the nearshore." The previous year, Chandra and Wittmann's 2013 study: *Development of a risk model to determine the expansion potential environmental impacts of Asian clams in Lake Tahoe* sharply contradicted TRPA and said that 200 acres of a potential 500 acres are infested. Even the lower number might be concerning if Tahoe's Asian clams caused any harm or threatened native mollusks, but they don't. Rubber mats on the lake floor harm Tahoe's ecology, not Asian clams.

Their final report to the Forest Service illustrates the lengths that Tahoe's aquatic researchers go to in order to protect their stature and diffuse criticism. Throughout, the report obscures results with irrelevant explanations and uses arcane language to describe simple procedures. It grandly concludes that even though the "ethylene propylene diene monomer barrier" (rubber mats) didn't work, "continuing to collect and provide this type of information will continue to support the integration of science and management at Lake Tahoe as a model for invasive species programs both nationally and globally."

The researchers report to the Forest Service was released in June 2013, eight months after the Emerald Bay project started, but nine months after repopulation surveys confirmed that Asian clam regrowth in the test areas was indeed exponential. When the researchers carpeted Emerald Bay with pond liner, they knew that there was no chance of controlling Asian clams in Lake Tahoe. When the matting was laid, TRPA said it would be in place for a year. Considering that four years earlier, TERC's Marion Wittmann had reported that after four weeks all of the clams were killed during the initial study, a year seemed an unnecessarily long time to smother the tiny crustaceans. But a test project at Emerald Bay had failed because strong currents had replenished oxygen under the mats, and a new technique using matting lined with decaying wood fiber to assist in oxygen removal was being used.

When the new technique was announced, it seemed that creating nitrogen-rich compost in Emerald Bay was an unorthodox way to decrease dissolved oxygen to kill clams. But regardless of the material used and how long the mats were left in place, it was known before they were laid that the clams would repopulate in two years, and it now appears that the change in material was a justification to proceed with the highly publicized commitment to control Asian clams in Tahoe's iconic bay—and to save face.

The project, of course, failed, and the eradication fiasco was explained away in the *Tahoe: State of the Lake Report 2014*: "The mouth of Emerald Bay presented unique habitat for clam survival and challenging conditions for the use of bottom barriers. High currents at the mouth tended to overturn barriers, and the porous and heterogeneous nature of the material at the mouth (a glacial moraine) allowed the flow of oxygen rich water through the sediment and under the barriers." This was exactly the same explanation given two years earlier to explain why the pilot test failed and why a different matting material would be tried.

The UC Davis report continues: "Despite the difficulties, a substantial reduction in clam density (90% mortality) was achieved across the treatment area based on sampling through October 2013." Using biodegradable mats achieved worse results than the test projects, and after leaving 35 tons of decaying mats in place for an additional year, the mortality rate dropped to less than 90 percent. No report on the Emerald Bay project explained why the researchers selected Lake Tahoe's preeminent tourist site for their first large-scale control project; one that had only minor Asian clam infestations and where a pilot project had failed.

Buried in the final Asian clam control report is the conclusion, "We recommend the continuation of a long-term experimental and monitoring program, [but without more research] it is difficult to understand the efficacy of a large scale [Asian clam] treatment program." (Wittmann, Chandra, Schladow, & Reuter, 2013) Said simply, we recommend continued funding to research and monitor a benign species that can't be controlled.

Sometimes new methods don't work. That wasn't TERC's failure; the error was in proceeding with a doomed project. When the Emerald Bay eradication program ended in November 2014, National Public Radio reported that Schladow said: "The surprise from the research is just how many clams are in the lake. You are talking about the highest concentrations seen anywhere in the world [are] at Lake Tahoe." Surely TERC's director knows that Lake Tahoe's Asian clam populations are some of the world's least concentrated, and unless he doesn't read the studies that carry his name, he also knows that it's not true when his colleagues say the clams are "responsible for unsightly algal blooms that wash up on shore and for calcium levels at which quagga thrive."

The control studies should never have been done and the projects shouldn't have been performed. The clams were known to have been in nearby Donner Lake for 40 years, and there they had caused no harm. Their biological functions and requirements were well known before they became an issue at Lake Tahoe. And the researchers knew—or should have known—that because Asian clam veligers free-float for a month before settling to the bottom, in open water, control by any method is impossible.

The Asian clam control program failures were glossed over and never publically acknowledged. The researchers' final report to the Forest Service makes it clear that it was wrong to malign the clams. Disproven science wasn't retracted or corrected, and at Tahoe it's still commonly believed that Asian clams are an ecological and economic threat to the lake, and TRPA repeats that message. In the fall of 2015 Aquatic Resources Program Manager Dennis Zabaglo announced that the mats would be rolled up and stored. He also confirmed that the agency remained undeterred in their war on Asian clams and wrote, "We plan to continue work in Emerald Bay, [but] an exact location hasn't been determined."

It would be difficult to contrive a more environmentally insensitive program than needlessly scouring five acres of Emerald Bay of most bottom-dwelling life by suffocating it under decaying matting, but TRPA's researchers may be devising one. The 2015 AIS implementation plan states that "the potential for pesticide treatment applied in combination with bottom barriers may provide a means for effective containment." The plan adds: "In scenarios where there may be a need to manage clams, such as where infestations may negatively affect water clarity, recreation, survival of native taxa or other important attributes of the lake, further investigation into various removal strategies to mitigate potential damages that Asian clams may cause is recommended."

At the 2015 AIS Forum, Zabaglo gave a preview of what the other strategies may be. He said that killing Asian clams was not only more complex than originally believed, but also required multiple methods. One of those methods was to use a "Zamboni-type machine" that he claimed was able to scrape the clams from the lake bottom and didn't harm the habitat. Considering that the clams are buried in up to a half-foot of sediment, it appeared that the machine was indeed miraculous if it could selectively remove them without seriously disturbing the dozen-plus native species that live on the lake floor. Another method being investigated is using electric rakes to kill the clams.

In October 2012, TRPA AIS Manager Ted Thayer said the mats would be in place for a year so as not to disrupt nesting egrets and other wildlife. In the March 2013 *TRPA Environment Report: Aquatic Invasive Species*, Thayer said the mats would be left in place through the summer, "so as not to interfere with high-boating season." Three months later, Tahoe RCD said, "it's to make sure all the clams die." Questioned about the efficacy and aesthetics of five acres of synthetic matting in Emerald Bay, Jeff Cowen, TRPA's public information officer said, "While a barge in the middle of Emerald Bay and large black mats on the lake floor are unsightly, the results will make for a better Tahoe."

After a year when the mats were still in place in Emerald Bay, TRPA was asked why they weren't removed as planned in the fall of 2013, Zabaglo replied, "The project is still in progress; monitoring is ongoing, and preliminary results are encouraging."

Previously, Thayer had said that because the mats were reusable, the high cost of the Emerald Bay project would be lessened in future projects. Apparently, the researchers hadn't told Zabaglo and Thayer that the clams would repopulate Emerald Bay in about the same amount of time as it took to kill them, or the TRPA managers hadn't read the researchers' reports that said the mats were ineffective, that the method was prohibitively expensive, and that control wasn't possible.

Researchers' reports to the U.S. Forest Service say that "all data has been discussed with members of the Lake Tahoe Asian Clam Working Group." In the spring of 2014, TRPA's Jeff Cowen responded to a request asking for the group's membership list and minutes from their meetings. He wrote: "No minutes are kept as this is an informal working group. We generally keep minutes only for meetings where policy decisions are made. The meetings are open to all stakeholders. Each interested organization is represented by appropriate staff members; there is no selection process for which organizations attend."

This raised the questions that if carpeting five acres of one of the world's scenic wonders with impervious matting at a cost of more than $1 million is not a policy decision, the public is not a Tahoe stakeholder, and environmental programs are monitored by an informal group of various agencies' staff at unpublished meetings, what other informal decisions do TRPA staff make?

Cowen also wrote: The mats "are part of a pilot study. The results they are achieving have yet to be determined." He too was apparently out of the information loop. Two months later, *Tahoe: State of the Lake Report 2014* reported that based on sampling in Emerald Bay through October 2013, 90 percent clam mortality had been achieved. A year earlier, regional media had reported that the Emerald Bay project was "the largest Asian clam eradication project in the lake's history [it was the largest in any lake's history]. The Lake Tahoe AIS program aims to eliminate a 5.5-acre population of Asian clams from the mouth of Emerald Bay and control the invasive species." The project was described as "a large-scale deployment of bottom barriers" to eradicate Asian clams. Even though there had been three previous "pilot" projects, after monitoring the project for a year, TRPA began calling the five-acre program a "pilot study," and after another year, said the mats would be stored with no plan for their reuse.

Six months after storing the mats, at the 2015 Tahoe AIS Forum, TRPA Aquatic Resources Program Manager Dennis Zabaglo boasted of the "success of the program to remove Asian clams from Emerald Bay." Zabaglo said TRPA had selected Emerald Bay to exterminate the clams in the five-acre program because the Asian clam population had exploded there in 2009 and that there had been a corresponding increase in algal growth. He explained that the "upgraded rubber tarps" had a layer of biodegradable fiber "intended to increase biochemical oxygen demand." What he didn't tell the Forum was that the upgraded mats achieved exactly worse results than those in the failed pilot program.

With a seemingly limitless flow of grant money, Tahoe's AIS scientists persist in their research to find an effective control method when even their own study states: "Asian clams' rapid recolonization rates likely pre-empts any potential for effective widespread removal and control in Lake Tahoe. This is supported by evidence from technical reports and peer-reviewed studies of Asian clam control experiments at Lake Tahoe and Lake George and the USGS Non-indigenous Aquatic Species database."

Soliciting contributions in April 2014, the same month that Cowen said the mats would be permanently stored, the influential Tahoe Fund proudly pointed to one of its four 2013 Signature Projects: Asian clam control in Emerald Bay. The bi-state non-profit had exceeded its fund raising goal and said: The Emerald Bay project "was the largest Asian clam control project in the lake's history, and the first time the technique, developed by UC Davis researchers, has ever been tried on this scale. The research team is now seeking funds to treat the remaining half-acre area of the population that was left untreated. Because the clam population is still small and isolated, the project could completely remove the remaining population from Emerald Bay and help control its spread to other areas."

A year before the Tahoe Fund's solicitation was made, Schladow told the media that complete eradication of Asian clams wasn't possible anywhere, TERC researchers knew that the clam beds would repopulate in about two years, and that the mats were ineffective and polluted the lake. And instead of using the Tahoe Fund's grant to kill more Asian clams in Emerald Bay, donor contributions to eradicate the clams were used to help offset the cost of rolling up and storing the mats—an operation that cost four times the Tahoe Fund's gift.

An explanation why Asian clams don't threaten native mollusks is given in *Commonly Rare and Rarely Common: Comparing Population Abundance of Invasive and Native Aquatic Species.* The comprehensive 2013 University of Wisconsin paper examined 24,000 invasive populations of 17 aquatic nonnatives compared to their native counterparts to test statistical patterns of abundance across multiple sites. The study concludes: "Our finding that AIS exist at low densities in most locations where they occur runs counter to the perception of invasive species as those that are abundant or dominant wherever they establish. ... Recognition of [non-uniformity] in densities introduces a new dimension to invasive species management."

The simple truth is that Asian clams cause no harm to Lake Tahoe, increase the lake's biodiversity, and improve water quality. In the NPR story in which Schladow incorrectly said Tahoe's clam concentrations were among the world's densest. TRPA's Tom Lotshaw said that the agency wants to continue the Asian clam control program but needs more money. He said Congressional passage of the Tahoe Restoration Act could help—$45 million would be allocated to AIS management and boat inspections—but TRPA needed public and private partnerships to help with raising the money to continue the programs.

Ignoring the expensive failures, TRPA put an unrealistically positive spin on the Asian clam control projects. In March 2015, Governing Board Chairman, Casey Beyer, wrote: "The Lake Tahoe AIS Program recently completed a pilot research project that used rubber mats to treat five acres of an Asian clam infestation at Emerald Bay. The goal of the project was two-fold: To evaluate the rubber mats' effectiveness as a way to reduce populations of Asian clams and other aquatic invasive species on a large scale at Lake Tahoe, and to reduce an Asian clam infestation at an iconic and heavily-visited recreation site." The rubber mats were ineffective; there were no other AIS populations reduced; many times more animals native to Tahoe had been killed than Asian clams; and the clams will repopulate to their original numbers in less time than the mats had carpeted Emerald Bay.

In the August 2015 *Implementation Plan for the Control of Aquatic Invasive Species within Lake Tahoe* Wittmann and Chandra placed Asian clams under the heading of "No feasible control options," and wrote that because of the clams' ability to quickly spread and recolonize that eradication or significant control wasn't likely. But in an appeal for research money said, "However, the potential for molluscicide treatment applied in combination with bottom barriers may provide a means to improve effectiveness of application and contact time. In scenarios where there may be an exceptional need to manage clams, such as where infestations may negatively affect water clarity, recreation, survival of native taxa [all premises the two researchers had debunked in *Natural and Human Limitations to Asian Clam Distribution and Recolonizations*], or other important attributes of the lake."

Notable in the August 2015 plan is that Chandra doesn't repeat his claim that Asian clam shells create calcium hot spots that facilitate quagga—the harm originally used to vilify the clams—and writes only in generalities about potential harms. The only specific harms Chandra and Wittmann potentially attribute to the clams are that "the Asian clam has been associated with (but isn't necessarily the cause of) filamentous algal blooms—[another remarkable admission after seven years of blaming Asian clams for causing algal blooms.] Through shell deposition, the Asian clam has had a negative impact on the aesthetic qualities of the nearshore environment in the lake."

Tahoe is certainly unique in that tiny shells on a sandy beach are considered by some aquatic researchers to be unaesthetic and hazardous. Sand Harbor is one of Lake Tahoe's iconic beaches. The AIS implementation plan emphasizes the elevated degree of Tahoe's Asian clam monitoring and the threats that their shells pose there: "In August 2014 at Sand Harbor, one half of an Asian clam shell was found on the beach about 50 meters south of the boat ramp. Scuba surveillance crews returned in September 2014 to do an exploratory survey off of the boat ramp and found one small, live Asian clam. During spring 2015 all suitable habitat within Sand Harbor State Park was surveyed for Asian clam and the shell from one recently deceased Asian clam was recovered [at 15 feet] off the boat ramp."

Steve Urie

— 8 —

The Elephant in the Room

The Tahoe Keys was built on one of the largest wetlands in the Lake Tahoe Basin. Research has shown that the wetlands of South Lake Tahoe used to remove tons of sediment and nutrients. The detrimental impact of this development can be easily seen during heavy runoff when plumes of sediment cause the waters to turn cloudy. ~ Tahoe Environmental Research Center, Docent Manual

TRPA did a remarkable job in shaping public attitude toward non-native aquatic species. However, their reports of nonnatives potentially inflicting hundreds of millions of dollars in damage, and of an impending catastrophic decline in Tahoe's tourist appeal because of shell-littered beaches strewn with rotting mollusk flesh, starving game fish, and globs of algae haven't happened—and won't—and it has nothing to do with TRPA's AIS prevention and control programs.

Lake Tahoe inspires awe, and after seeing it, most become protective of it. That's why in 1997 President Bill Clinton promised "to protect Lake Tahoe's environment, and with it the area's economy and quality of life." Since then state and federal government has funded $100 million a year for its protection. Annually, senators and governors return to renew the pledge to protect Tahoe and preserve its grandeur. Their commitment is passionately shared by thousands of the Basin's residents and visitors. Occasionally, those who are responsible to prevent it from harm become overly zealous, exaggerate threats, and distort science to support opinions and programs.

TRPA's persistent message that the lake is at peril from AIS has been internalized by the Basin's residents, and well-meaning citizens passionately take TRPA's protective directives to heart. Expanding on grim misinformation from TRPA and Tahoe RCD, after the 2012 boating season an Incline Village resident wrote in the *North Lake Tahoe Bonanza*: "With the boating season winding down and most watercraft being removed from Lake Tahoe before winter storms arrive, it's an excellent time to contemplate whether the time has finally come to move swiftly in the direction of regulations that will prevent invasive species such as zebra and quagga mussels from taking hold in the lake environment.

"The Tahoe economy will suffer greatly and property values will drop precipitously if the lake becomes infested with aquatic invasive species multiplying out of control. We cannot allow a recreational pursuit to destroy property values all around the lake simply because a few out-of-area boaters are careless or unconcerned about decontaminating their watercraft properly.

"While out-of-area boaters may complain about a stringent new regulation ... it would be very easy to implement a program where 'Tahoe Only Boats' are licensed and registered and stored in approved facilities. ... It's not fair to expect Tahoe property owners to risk economic devastation for the selfish enjoyment of a small group of non-residents."

When passion is honest and doesn't infringe on others' rights, committed people can be forgiven uninformed rhetorical excess—especially when their opinions are shaped by generally trusted sources. Agency managers, researchers, and academics don't get the same break. Their work is largely supported by public money, and because often their statements are on complex subjects, of necessity, they receive blind public trust. And because they drive public policy, they are held to higher standards.

It's why on a February 2013 morning some cringed and others sadly shook their heads when they heard Al Roker express mock shock and disbelief on NBC's Today Show when Willie Geist exclaimed giant goldfish were lurking in Lake Tahoe. Roker is an experienced newscaster and journalist who knows alarmist news from substantive issues. The report led with a four-year-old picture from TERC's files of a UNR researcher using the old fisherman's trick of holding their fish close to the camera lens to exaggerate its size, and Today Show reporter, Kristen Dahlgren explained how as the lake has gotten warmer, "goldfish were thriving." The reality is only a couple of dozen have been found.

Back-dropped by Tahoe's snow-capped peaks, shocked UNR researchers and a somber TRPA manager told the national TV audience that monster goldfish were threatening Lake Tahoe. Sudeep Chandra quizzically posed a rhetorical question, "We know we have a giant goldfish, the question now is 'how long has it been there?" For a limnologist, who extensively studied Lake Tahoe, he should know the answer—any Tahoe Keys resident could have told him: maybe as long as 20 years. He then asked, "And how many others are in the lake?" He should know the answer to that question too: none. No Tahoe Basin goldfish has been found in the wild outside of the Tahoe Keys' manmade habitat behind the lake's shoreline, and as Chandra has pointed out, the Keys aren't even in Lake Tahoe. But the larger question was: why fabricate an issue that distracts from Tahoe's important environmental concerns?

The stale, resurrected story of Tahoe's giant goldfish was carried by news media ranging from the Associated Press and Reuter's news services to *Unexplained Mysteries*. UNR's *Nevada*

Today reported, "The invading goldfish have sparked international interest and considerable media coverage around the country, including a segment on NBC's Today Show. For Chandra and his team, the media coverage is welcome. It highlights attention on this important environmental issue and the successes of the researchers and faculty in the Aquatic Ecosystems Analysis Lab." No successes were mentioned, and many believe there are environmental issues facing Tahoe that are more deserving of international attention and TRPA's Communications and External Affairs staff time than designing misleading stories to bolster support for UNR's aquatic ecosystems lab and TRPA AIS programs.

The Tahoe Keys was built in 1959, a decade before TRPA was formed and before the importance of Tahoe's wetlands to its aquatic habitats and the lake's water quality and clarity were fully understood. The shore-side real estate development turned an apparently useless 1,300 acre wetlands at the mouth of the Upper Truckee River into a 740-acre marina-community. By excavating 5-million tons of marshland, 11 miles of shallow canals were dredged, allowing lake access for boats from 1,500 home sites.

Today, five miles north of the Keys at the Taylor Creek Stream Profile, U.S. Forest Service educational materials explain how over millennia wetlands are formed by sediment settling at the mouths of streams flowing into Lake Tahoe. As sand, silt, and decomposing vegetation deposits are accumulated, hydric soil is formed, creating a habitat suitable for aquatic plants that help to further filter stream runoff and add to the soil. The natural cycle that built Tahoe's wetlands constructed a filtration and water purification system that worked exceptionally well for thousands of years. The lake's water is still 99.994 percent pure, making it one of the purest lakes in the world—commercially distilled water is 99.998 percent pure.

In the early years after the destruction of more than a square mile of wetlands at the mouth of the Upper Truckee River, Lake Tahoe was able to absorb the river's flow without noticeable effects. And before homes were built and lawns and parkways planted, the Keys' canals and lagoons mirrored the clear water beyond the river's engineered, twin channels that allow seepage and overflow from the river to flow into Tahoe. But as marshland soil was enriched by runoff loaded with fertilizer, plants began to grow in the warm, stagnant waterways and algae increasingly clouded the canals. Today by mid-summer, particulate-saturated, algae-thick water flows into Lake Tahoe, and a quarter-mile-wide, greenish-beige band wraps along South Shore beaches.

Sixty-three streams flow into Lake Tahoe, and wetlands at the mouth of most protect the lake's famed clarity. But a quarter of Tahoe's inflow is from a single source, the Upper Truckee River. After flowing 23 miles from Grass Lake and collecting outflow from 15 lakes and streams, the river flows through less than half of the former marsh and adds it's sediment to the murky Keys' canal water before mixing with Tahoe's crystalline waters.

The loss of 60 percent of the Upper Truckee Marsh in order to build the Tahoe Keys punched a hole in nature's filter for sediment and created a warm-water bayou capable of supporting a variety of aquatic life found nowhere else in the High Sierra. The marsh's destruction is the single largest contributor to the decline in the lake's clarity. And unlike Lake Tahoe's sand, gravel, and rock bottom, the Keys rich hydric soil is hospitable for aquatic plants, and it has become a refuge and preserve for a half-dozen non-native fish that can't establish in the lake. In his paper *Distribution and impacts of warm water invasive fish in Lake Tahoe,* Sudeep Chandra says that the Tahoe Keys are outside of the lake, a critical distinction to be noted when projecting the impact of Keys-bred non-native plants and animals on all of Lake Tahoe.

Meaningful projects by the California Tahoe Conservancy and the Forest Service are restoring long stretches of the Upper Truckee River and Marsh and mitigating the on-going toll levied by the Tahoe Keys, and TRPA has begun to buy back and trade land for wetlands restoration. But the agencies can never restore the protection that was once provided by the marsh. TRPA justifiably boasts that in the Environmental Improvement Program's first decade they "restored more than 739 acres of wetlands" in the Tahoe Basin—ironically, almost exactly the same amount of wetlands destroyed by the Tahoe Keys.

A persuasive argument can be made that any plant that takes root in the Keys is simply nature trying to reclaim what mankind took and contributes more to the lake's natural ecology than trying to control it could ever do. From the 1960's until the early 1980's, nature's efforts to reclaim its lost wetlands were hardly noticed. But as the canals turned cloudier and plants began to snag boat propellers, homeowners and vacationers became upset. Likewise, naturalists were offended when rainstorm and spring runoff wrapped algae-packed plumes of silt and fine sediment from the Keys along the South Shore.

During the 1990's, aquatic plants in the Keys became a problem, clogging the canals and making once-clear water aesthetically unappealing. As distressing to homeowners was that nature was fighting back, and her efforts were forcing expensive summer-long weed harvesting, so that boats could navigate from backyard docks to the lake. And unwittingly, a homeowner or two had given nature a helping hand by dumping aquariums containing Eurasian watermilfoil, a hardy fast-growing aquatic plant, into the warm, hospitable canals behind their homes.

The *Tahoe Daily Tribune* reported that by 2001 the Keys were: "being choked by an aquatic weed and little [was] being done to stop its costly, invasive spread. The Lahontan Regional

Water Quality Control Board, which oversees the lake's [water] quality control, [had] for several months studied the effects of Eurasian watermilfoil on Lake Tahoe. In the meantime, the [plant had] more than quadrupled [in the Tahoe Keys and] cost property owners, boaters, and marinas hundreds of thousands of dollars in control and corrective measures.

"Despite predictions of huge financial loss, drastic changes in the biological makeup of the lake, and imminent mass spread of the hardy plant, the local water quality agency placed a non-priority label on the issue." Fears of financial loss, a drastic change in lake Tahoe's ecology, and wide spreading of watermilfoil didn't materialize. But within four years, UNR and TERC researchers had teamed up with Tahoe RCD to combat the non-native plant. The researchers said, "Once established, [watermilfoil] degrades fish habitat, fouls boat propellers, chokes out native plants, and accelerates algal growth—except for fouling boat props, the rest of the list of watermilfoil complaints are over-stated or untrue.

Tahoe RCD took their battle against aquatic invasive plants to Emerald Bay, where Eurasian watermilfoil was first discovered in 1995. By 2003 a survey found that it had expanded to over an acre along the western edge of the bay. And two years later, a pilot project to remove the plants was initiated by Tahoe RCD with assistance from TERC and UNR. It was hoped that divers could eliminate the plants by hand-picking them. A July 2005 *Tahoe Daily Tribune* article reported that for three days "a dive crew removed milfoil from Emerald Bay. One diver pulled plants, and a boat crew collected the plants and fragments with a dip net. They were able to remove 65 pounds of fresh weight milfoil." The team said the "removal was successful" and planned to return in two months "to measure and weigh any milfoil that they missed."

The watermilfoil immediately returned, and over the next four years, a series of small-scale treatments were deployed, but the infestation continued to spread, and by 2009 three patches of watermilfoil covered a combined area of nearly three acres. The following summer, Tahoe RCD reported, "Participating managers are implementing a strategy for the effective removal of aquatic invasive weeds based on risk of spread, public benefit, and the likelihood of removal success. The swimming beach and pier area in Emerald Bay was identified for the pilot project because it meets all of these criteria."

In 2010, a decade after announcing the first successful eradication Emerald Bay's watermilfoil, Tahoe RCD began a 5-year program to remove 3.6 acres of weeds from the bay. Three years into the program, in the spring of 2014 only 12 Eurasian watermilfoil plants were found and removed from Emerald Bay. No plants were later found in a fall survey, and the bay was again declared watermilfoil free. At the 2014 Tahoe AIS Forum, TRPA biologist Patrick Stone said that because watermilfoil can grow from stem fragments, removing watermilfoil from Emerald Bay would be a continuous process, and that the plant will always be in Lake Tahoe. In August 2015, Dan Shaw, an environmental scientist with California State Parks, said that even though he still returns to eliminate new watermilfoil growth as it sprouts, Emerald Bay "was thoroughly purged of the persistent weed."

Tahoe Keys' inlets, lagoons, and canals wind through 172 acres of the former marsh, creating a veritable aquatic Garden of Eden. Tahoe has eight native aquatic plants and two nonnatives: Eurasian watermilfoil and curlyleaf pondweed. All are found in the Keys. Whereas all 172 acres of the Keys' waterways have dense aquatic plants, only about 80 acres of Lake Tahoe's 122,000 acres of lake bottom support aquatic plants. At the forum, Stone said that other than the Keys, watermilfoil is only in Emerald Bay and some areas where the bottom was dredged for piers and marinas.

TRPA accurately says, "Once established, invasive plants help establish 'warm-water fish' and amphibians in Lake Tahoe." Before the Keys was dredged, there were, of course, no fish there, but because nature abhors a vacuum that quickly changed. Adaptively suited for living in the lake's ultra-oligotrophic water, few native Tahoe fish are found in the warm lagoons, but goldfish from aquariums and minnows dumped from bait buckets are thankful to live out their years in the hospitable ecosystem.

Many believe fish evenly distribute throughout a lake. Anglers know that is untrue—fish go where food is most plentiful and the habitat is protective. For fish to sustain in any ecosystem there has to be enough food and cover. The Keys offers a fine habitat for warm-water fish, and not everyone views them negatively. Long before the Today Show's monster goldfish alert, a vacation rental website boasted, "Keys lagoons are filled with bass, crappie, trout, and even enormous goldfish. Fishing from a small boat or private dock gives you the chance of catching a Keys Grand Slam (rainbow, brown, crappie, bass, and bluegill)"—all non-native to Lake Tahoe. While some are proud of the fine fishing in the Keys, others are trying to ruin their fun. In 2011, UNR researchers initiated an aggressive program to remove warm-water fish from the Keys. In the first year, more than 12,000 were caught by gill-netting and electrofishing.

But the Keys major contribution to Tahoe anglers is that, pushed by flow from the Upper Truckee River, its plankton and algae-rich water supplies a steady food source for mysid shrimp and crayfish, which in turn feed Tahoe's game fish. Although mackinaw are hooked in the deep water off the Tahoe City shelf and Cal-Neva Point, Tahoe's best sport fishing is in the nutrient-rich stream that flows east from the Keys along the South Shore ledge to the Logan Shoals area. There, brown, rainbow, kokanee, and record-size mackinaw are caught year-round. Crayfish and mysid shrimp are considered invasive by TRPA, but researchers have determined that they can't be controlled.

A 2009 study by UNR, the Forest Service, and Tahoe RCD justifies TRPA's warm-water fish eradication program. The *Tahoe Daily Tribune* described it: "The study is a follow-up to a pilot project last year that showed warm-water fish species were able to leave the relatively warm refuge of the Tahoe Keys for other parts of the lake, Sudeep Chandra said. In the latest study, bass were stunned using electroshocking equipment and brought on shore, where researchers sedated them using club soda. Researchers then made a small incision on the bass's bellies to implant two tags, before stitching the fish back up and returning them alive near the area [where] they were caught"—an incredibly sensitive procedure, considering that two years later the same researchers were killing bass by the thousands.

"By studying bass movements past 13 receivers that have been placed in the Keys during the next two years to pick up signals from the acoustic tags, researchers should be able to determine how often bass leave the Keys and gain greater knowledge about what triggers them to do so, Chandra said.

"Chandra expressed doubts about whether warm-water fish species could ever be eliminated from Lake Tahoe, but said this study will be a critical step to figuring out what to do next. ... Because the species were recently introduced to the lake, their spread can likely be controlled. Using electrofishing to remove fish from the lake and eliminating the dense underwater forests of milfoil and pondweed where the warm-water fish thrive are two ways the fish populations could potentially be limited. The warm-water fish removal program primarily targets fish only in the Tahoe Keys where they are caught by electrofishing."

The researcher's premise was wrong, the reason almost all bass are in the Keys is because it's the only area with suitable habitat. And although no reports were given on how often the fish with tracking tags in their bellies left the Keys, or where they went—anecdotally, bass fishing in Lake Tahoe is still terrible.

A *Reno Gazette-Journal* article described Chandra's study, the history of fishery management at Lake Tahoe, and the damage non-native fish supposedly do: "Largemouth bass and bluegill are thriving around Lake Tahoe, a situation that could be accelerated by a warming climate. Those are among the conclusions of a new study of warm-water fish that live in the Sierra lake and the threats they could pose to its sensitive ecosystem. 'The most important findings from our study indicate that the warm-water fishes of Lake Tahoe are either competing (with) or preying upon native fishes,' said Sudeep Chandra.

"The good news, Chandra said, is that densities are low in most locations, [and] if steps are taken now to manage those fish, changes may be possible 'before they get out of control.' Recently introduced [bass and bluegill] bring their own problems. In the Tahoe Keys, where temperatures are warm enough for the fish to spawn, native minnows have virtually vanished.

"'Until we get a good handle on this, it's hard to determine the best way to move forward.' If the fish proliferate in number, they could affect Tahoe's diminishing clarity by boosting algal growth with nutrients excreted into the water,' Chandra said. Largemouth bass eat crayfish, which were introduced into Tahoe and established there by 1936. The lake's crayfish, which he said may now number more than 200 million, contain large quantities of phosphorus. When the bass eat and excrete crayfish, the released phosphorus could fuel additional algal growth in Tahoe.

"The large number of crayfish in Tahoe could provide a ready, untapped food source for bass that could help boost their numbers, experts said. 'If we do not begin to manage populations, especially source populations such as the Tahoe Keys, then it is likely their densities will increase and result in more stress to Lake Tahoe's native biodiversity and impact clarity,' Chandra said."

Chandra's media statements raise a number of questions: The Keys were dredged a half-century ago, and "native fish" are those that were in a habitat prior to human influences; how can native minnows virtually vanish when there weren't any? And because mackinaw and rainbows played major roles in eliminating Tahoe's only native trout, shouldn't they be held in as much contempt as bass and bluegill? And if bass are environmentally destructive, why were they introduced into Martis Lake, which is only five miles north of Lake Tahoe? And compared to the algae and sediment that continuously flows from the Tahoe Keys, does anyone believe the Keys' warm-water fish are even a minor factor in reducing the lake's clarity? And since almost all of the predator-prey/native-nonnative discussion is limited to the Keys, instead of letting kids catch them, why are fish being zapped by researchers?

After the first year of the Tahoe Keys warm-water fish control program, TRPA's Ted Thayer said that despite the ton of fish that were removed, "capture rates didn't go down, so it is doubtful the removal had any impact." When asked what was done with the 12,000 fish that were killed, Thayer said they "were taken to UNR for research." Four years later in August 2015 after killing more than six tons of Tahoe Keys fish, apparently fewer were needed for research, and a UNR press released boasted that 200 pounds of fish from the Tahoe Keys had been donated to a local food bank.

As Thayer noted after the program's first year, killing tens of thousands of fish only temporarily reduces their populations, and hopefully UNR researchers and TRPA will learn that in a hospitable habitat, they can't kill fish faster than they reproduce. Of Tahoe's seven native fish, all but the Lahontan cutthroat trout remain, and none of the natives are endangered. Wildlife services have introduced five sport fish into the lake, so isn't it gratuitous for the researchers to say bass and bluegill in the Keys stress the lake's native biodiversity? But larger questions are: since the Tahoe Keys is not part of Lake Tahoe, why are the ecological issues of its warm, artificial ecosystem extrapolated onto the lake as a whole, and why does TRPA spend public money to mitigate environmental problems caused by the private development.

At the same 2014 assemblage of journalists where the success of their warm-water fish control program was explained, UNR and TERC researchers also promoted the need for more funding in order to monitor and gather data to better understand the lake and to "convince our representatives and stakeholders to invest in the monitoring program." Some believe a redirection of funds from non-productive AIS prevention and control programs may be a start to solving Tahoe's aquatic research funding crisis.

Chandra says zapping bass reduces their numbers. That's indisputable, but it's only temporary, and he should take a lesson from outdoorsmen. Two-time Outdoor Writer of the Year and America's first Back Country Sportsman of the Year, Tom Stienstra who points out: "It doesn't work that way. The amount of fish or wildlife that a habitat can support is determined by the amount of water, food, and cover that is available. It is called carrying capacity. A habitat with low carrying capacity means few fish. It is habitat that determines population numbers. The first rule is that fish and wildlife need habitat to survive." (Stienstra, 2014)

In his book, *California Fishing,* Stienstra explains why Lake Tahoe's purity and rocky substrate undermine its ability to support large populations of aquatic life. Stienstra writes that due to its lack of vegetation and cover, 95 percent of Tahoe's nearshore has extremely low carrying capacity and supports no fish. But in the Keys, as its lagoons and canals become murkier, their carrying capacity will increase and the fishing will get even better. It's why zapping fish in the Keys is a waste of time.

In his 2015 AIS control plan, Chandra indicated that he is still unclear on the overriding importance that the Keys' artificial habitat plays in supporting warm-water fish. He wrote: "Warm water fishes such as bass and other species were first discovered in Lake Tahoe in the 1970s and 1980s. The populations of these fishes are increasing through time, with the majority of the populations concentrated in the Tahoe Keys area of the lake."

Overlooking the fact that Tahoe's native fish are in the lake, and virtually all of the nonnatives are in the Keys, Chandra said: "Warm water fishes prey on native fishes, and contribute to the ongoing decline in native species. In Lake Tahoe, since 1960, there has been a tenfold decline in native fishes. We've found that we must target the large spawning [non-native] adult fish, and need continuous efforts, on a yearly basis, to be effective"—an unsubstantiated statistic and totally inaccurate assessment.

Tahoe's aquatic environmental objective should be to minimize unnatural habitats, not the animals that live in them. And instead of sinking millions into AIS prevention and control programs, including the futile slaughter of fish in the Tahoe Keys, if that money were redirected to wetlands restoration and runoff management, there would be meaningful progress toward improving the lake's nearshore water quality and clarity.

In 2009 the *Lake Tahoe Region AIS Management Plan* reported that "at least 20 non-native species are established in the Lake Tahoe Region, including aquatic plants, fishes, invertebrates, and an amphibian." The plan says, "Eurasian watermilfoil has been spreading around Tahoe, and curlyleaf pondweed has begun to expand dramatically. Beds of Asian clams are larger and more common than previously known, and populations of warm-water fishes such as largemouth bass and bluegill are expanding"—all greatly exaggerated claims.

It's comforting when a program is implemented to protect the environment. It usually demonstrates that without political implications, government is protecting the public interest. Even though the Asian clam control program wasted millions of dollars and caused incalculable ecological damage, no one was harmed. Likewise, except to the fish, there is no harm in zapping them. So, no one complains about the expensive ongoing slaughter of fish in the Key's, and Tahoe's warm-water fish control program remains low on the list of public programs in need of reform.

When control programs such as poisoning lakes for aquatic invasives fail, they are usually abandoned when it is realized that more harm was done than benefit was received. After DDT ushered in the "pesticide era" in the 1940's, the chemical's Swiss discoverer was awarded a Nobel Prize. Twenty years later, the publication of *Silent Spring* exposed that DDT was not only killing fruit and vegetable-eating bugs, but also decimating birds such as the peregrine falcon and bald eagle.

A decade later in California, a Medfly infestation was attacked by government helicopters spraying Malathion. The insecticide isn't harmful to humans, but in 1989 scientists declared that it was impossible to eradicate the agricultural pests. Instead of abandoning the program, the state intensified it, often spraying each new infestation more than a dozen times. In one year alone, the escalated program cost $120 million—and produced no measurable results.

Except at stream mouths and where bottom sediment has been disturbed by dredging, few aquatic plants sprout from Lake Tahoe's rock-strewn, gravelly bottom, but in the Keys' nutrient-rich waterways, the volume of aquatic weeds has exploded. In 2007, 4,400 cubic yards of watermilfoil were harvested. By 2014 the volume of plants quadrupled to a volume that filled 1,800 dumpsters and weighed 350 tons. At the 2015 Tahoe AIS Forum, Rick Lind, President of Sierra Ecosystems told the assemblage of agency managers that mechanical harvesters—floating mowers—were no longer effective and herbicides were being considered. Two months later, the Tahoe Keys Property Owners Association announced a "science-based plan" to use chemical herbicides to control the prolific growth of curlyleaf pondweed, Eurasian watermilfoil, and coontail—a hardy plant native to Lake Tahoe.

Retired U.S. Department of Agriculture invasive weed specialist Lars Anderson headed a $250,000 study for the TKPOA that was reviewed by five Tahoe area scientists. The panel of experts proposed "tailored and prescriptive use of approved herbicides" to attack the plants choking the Keys' waterways. Acknowledging that they couldn't be eradicated, the scientists said the weeds could be chemically controlled.

Anderson said that the program's goals were to reduce the $400,000 annual cost to mechanically harvest the plants and "to improve the habitat for native plants and fish that's also useful for humans to sail, and swim, and boat in"—an ambitious goal that didn't address the issue that if selective herbicides were effective, reducing the two nonnatives would allow coontail and other natives to flourish in their place. The Keys' shallow, warm water and soft-sediment substrate is ideal for aquatic perennials.

Herbicides are prohibited in Lake Tahoe, but the EPA allows the Lahontan Regional Water Quality Control Board to issue exceptions in special cases. When asked about human health issues, TKPOA president John Larson said if contaminants were found in the drinking water, one of the wells operated by the TKPOA Water Company could be taken off-line, and in the worst-case, the Keys could get its water from the South Tahoe Public Utility District.

California Department of Fish and Wildlife scientist Joel Trumbo, one of the experts who reviewed the plan, told environmental journalist Ben Goldfarb. "Herbicides are directed at plant physiological processes. The potential herbicides they'd be using on this project run from slightly toxic to fish down to practically non-toxic" to humans (Goldfarb, 2015). At the plan's announcement, the panel of scientists told the public, "Herbicides are toxic to plants. The converse is they are less risky to things that aren't plants."

It was those and similarly dubious qualifications and ambiguities, and the subjugation of environmental and human health concerns to homeowners' amenity, aesthetics, and cost savings, underscored by Chandra's statement that herbicides are "more efficient and cheaper," that turned most Basin residents against the plan. Judith Michaels Simon wrote in *Moonshine Ink*, "Please don't pat me on the head and assert that a little herbicide won't harm me," and argued that "poisoning our lake for the convenience of homeowners isn't the answer."

Earlier the same week that the Keys' weed management plan was presented, Wittmann and Chandra's implementation plan to control Tahoe AIS was released. The researchers wrote: "While highly controversial, herbicide treatments may provide a cost-effective means to reduce local infestations and limit the spread of watermilfoil. Disadvantages include restrictions to swimming, drinking water, and fishing, and potential impacts to [native] plants. Additionally, the use of chemical controls may require extensive water quality monitoring."

TRPA Executive Director Joanne Marchetta wrote of the AIS implementation plan: "It identifies the species we have the best chance to control or possibly eradicate. ... The plan points to the control of invasive weeds Eurasian watermilfoil and curlyleaf pondweed. These are harmful species we can efficiently control or eradicate with the right projects in the right places." Most Tahoe Basin residents believe that weeds in the Keys canals are less invasive than toxic chemicals.

Before it was announced, the Tahoe Keys' herbicide plan had been in motion for more than two years. Sierra Ecosystems, a firm that specializes in pest management and land planning, permitting, and development contracted with Anderson to study possible solutions. The agencies whose approvals were needed to allow herbicides were engaged, and a month after the Keys' plan was released, the U.S. EPA allowed the Lahontan Regional Water Quality Control Board to approve herbicide use on a case-by-case basis in the Tahoe Basin—the wheels were set in motion to use toxic chemicals to kill plants in the Tahoe Keys.

Intuitively, it's a bad idea to introduce toxic chemicals into one of the world's purest bodies of water. And similar to the ill-fated Asian clam control program, a desire to satisfy the aesthetic and environmental whims of a few was being advanced without a thorough understanding of potential harms. Also like the Asian clam program that preceded it, the Key's weed management plan is experimental—none of the proposed "approved herbicides" have been successfully used in a similar situation, and all come with the possibility of doing more harm than good.

In its draft form the plan seems reasonably constructed. In 2017, the first year of herbicide usage, only a few small areas will be treated, and they will be closely monitored. By the third year, it is anticipated that almost half of the Keys' canals will be treated, and by the fourth year, the non-native weeds are projected to be under control and mechanical harvesters won't be needed. It is also optimistically projected that after year four, there will be a significant reduction in the need for herbicide treatments.

The expected outcomes are illogical. From 2011 to 2013 the TKPOA tested light-blocking bottom barriers to control the weeds. The plan says the "study report stated that nuisance aquatic plant growth was suppressed in the short term (less than one year) but recolonization over the subsequent two growing seasons resulted in plant densities similar to what was found in untreated areas." Because the areas treated with herbicides will be staged, there will always be untreated areas in the Keys, and the nuisance plants will regrow the following season.

One of the negative byproducts of killing the plants with herbicides is that unlike harvesting the plants and taking them to landfill, the plants killed chemically will sink to the bottom and add to the unsightly, nutrient-rich muck that is already there, and will assist future plant growth. The science-based plan takes a narrow perspective of the regenerative nature of a hospitable habitat and assuredly guarantees the plan's ultimate failure.

TRPA Executive Director Joanne Marchetta wrote of the Keys weed management plan: "We applaud these [Tahoe Keys] property owners for starting this critically important process to address the issue. Doing nothing is not an option." Marchetta mischaracterizes the issue—virtually all Tahoe visitors and non-Tahoe Keys residents believe that doing nothing is a much better option than using toxic chemical herbicides to kill the weeds.

However like the Asian clam control program, after the five-year program ends most will have forgotten what the researchers' initial reasons and projections were. And similar to the AIS scare promulgated by Tahoe agencies, disinformation disguised as environmental outreach will be fed to the public. One of the more effective information programs is an age-appropriate AIS curriculum growing from one elementary school year to the next. An October 2015 *Tahoe Daily News* article reported that Forest Service instructor Adilene Negrete told one class: "When you have invasive species, it hurts native species," and the League to Save Lake Tahoe's Savannah Rudroff told the class that Eurasian watermilfoil and curlyleaf pondweed are two of the worst nonnatives. The article didn't say whether the students were told of the possible harms that the chemical herbicides posed to their health and to native fish and plants.

After an environmental protection program is budgeted from public funds, it usually becomes entrenched. Such a program is familiar to Tahoe residents and visitors. In the early 1920's at a time when entomologists were still identifying California's native flora and fauna and decades before the invention of pesticides, the California Department of Food and Agriculture thought it prudent to assure that vehicles entering the state weren't transporting bug-infested plants or produce. Inspection stations were erected on all highways leading into the state, ironically to prevent insects from entering where most already were. For example, the mountain pine or bark beetle has been in California as long as anyone can remember, but cords of wood are still confiscated at the state line because they carry beetle markings.

When budget constraints forced the closing of 11 of 16 inspection stations in January 2004, many breathed a sigh of relief and hoped that the anachronistic program would die from lack of funding. But when the budget crisis ended, the inspection stations reopened in July 2007. Similar to no new aquatic invasive animals being identified in any of the Sierra Nevada's 150 large lakes during Tahoe's first seven years of boat inspections—no new bugs invaded California during the three years the vehicle inspection stations were closed. Today, as motorists waste time and gas as they creep past bored ag inspectors at the state line stations, most wonder about how an intuitively needless program continues to be funded.

No one intends to harm the environment, but occasionally ignorance of the science or an over-exuberance to restore what was does just that. In less than a decade, more than $30 million dollars was spent on AIS protection and control at Tahoe. Yet not a single dollar of financial harm was attributed to the lake's non-native aquatic plants and animals. And there is overwhelming scientific evidence that none of the aquatic animals that boat inspectors are hired to prevent from entering Lake Tahoe can establish there. And instead of spending less money than Tahoe boaters pay each year in inspection fees to verify that the lake is truly at risk from an aquatic species, without evidence TRPA vilifies plants and animals that pose no risk to Lake Tahoe and lobbies for public financial support to continue a baseless boat inspection program.

Steve Urie

— 9 —

Lake George: Bad Advice from Lake Tahoe

In 2002 researchers identified invasive Asian clams in the southern part of Lake Tahoe, but no action was taken. By 2008 Asian Clam beds and algal blooms had turned the once crystal clear water green. ~ Emily DeBolt, Director of Education, Lake George Association

How did the Director of Education for the Lake George Association get the idea that Lake Tahoe had turned green? She read it in the Executive Summary of the *Asian Clam of Lake Tahoe: Preliminary scientific findings in support of a management plan,* co-authored by UC Davis and UNR professors. On the paper's first page it is reported: "In 2008, nuisance-level green algal blooms correlated with Asian clam presence ... impacted water clarity, beach aesthetics and swimming areas." (Wittmann, et al., 2008)

The greening of Lake Tahoe rumor had actually emerged the year before when a March 2008 NBC News story said that a study by TERC directors Geoffrey Schladow and John Reuter concluded that water temperature "changes could turn Tahoe's famed cobalt-blue waters to a murky green in about a decade" And Schladow said, "A permanently stratified Lake Tahoe becomes just like any other lake. It is no longer this unique, effervescent jewel, the finest example of nature's grandeur."

Seven years later, using hyperspectral radiometers, TERC researcher Shohei Watanabe disproved Schladow and Reuter's prediction that Lake Tahoe was turning green and demonstrated that from 2012 to 2015 the lake became increasingly bluer. Watanabe's research also disproved Sudeep Chandra's visual observations. In August 2014 Chandra said, "Tahoe seems to have a greenish hue this year." (As measured by Watanabe, it was actually bluer than the previous two years.) A month after saying that Lake Tahoe was greener, Chandra took his observational pseudoscience to the nation's only lake that is clearer than Tahoe.

National Public Radio reported: "Chandra says like Tahoe, crayfish numbers in Crater Lake are also increasing. In Crater Lake they've doubled in the past five years. Chandra says that is one of many factors that could be affecting the lake's clarity. 'We have been finding the edge of the lake turning slightly green. Is that because of a changing climate, drought, or is it because crayfish are excreting nutrients at a higher rate causing algae to grow.'"

Chandra also blames crayfish that were introduced in 1914 by Crater Lake's park superintendent to help feed trout for preying on the lake's native Mazama newts. For a century, the crayfish and native newts maintained an ecological balance, but then in 2012 park biologists declared that crayfish populations were exploding and that they were decimating the newts. Park officials are seeking endangered status for the newts, and they called in Chandra to lead a team of scientists to determine the best way to control the crayfish and protect the newts.

Saying that "in an ecosystem of this size and complexity, it's not likely that crayfish can ever be eradicated," Chandra advocated an experimental control plan and said that the "only recourse is to keep the invasion front from expanding." Similar to his claims about Asian clams, Chandra's claim that crayfish waste may cause algal growth is wrong. Crayfish eat detritus, carrion, and even their own molted shells, and they remove significantly more polluting biomass from the water than they add.

Chandra concluded that "One would think at a beautiful lake like Crater Lake, with nice blue water with very little algae, you'd want to keep algae away." (Goldfarb, 2015) To accomplish that, he and his team of scientists proposed building aluminum walls to isolate the crayfish from shoreline areas that over the last century the resilient crustaceans haven't colonized. Considering that crayfish are found in Crater Lake at depths of 800 feet, the crayfish walls may surpass the attempt to control Asian clams with rubber mats as the world's most absurd AIS control plan.

In the summer of 2015, Chandra received $400,000 from the Lahontan Regional Water Quality Board and the Geological Survey for a ten-month study to determine why algae are increasing in Lake Tahoe. Chandra said: "This is a unique study; we're not just monitoring the lake, but looking at the mechanisms causing the algae to increase." Based on previous conclusions, it is predictable that Chandra will blame crayfish and Asian clams—two species that greatly reduce algae—and ignore the application of fertilizers by humans, a species that greatly increases algae.

In August 2010, Asian clams were found in Lake George. The Lake George Association formed a task force, and reported that the following month "scientists from Lake Tahoe visited to share their experience with fighting Asian clams." The visiting scientists were Sudeep Chandra and Marion Wittmann, and TRPA's Ted Thayer accompanied the researchers to New York. Chandra told the task force, "Other ill-effects associated with the Asian clam include its tendency to ease the introduction of other invasive species. The dead shells become a substrate for invasive mussels. The shells also produce calcium, which alters the water chemistry to the advantage of the mussels."

Lake George's clam beds were denser than Tahoe's, but it was believed they were contained in a 4-acre area. Modelling their program on TRPA's, in April 2011 the task force released a *Plan to Contain and Eradicate the Infestation of the Invasive Species Asian Clam in Lake George*. The plan states: "The most comparable research to Lake George comes from Wittmann, et al. After monitoring the clam's exceptional range expansion over the course of eight years in Lake Tahoe, Wittmann suggests the potential economic and ecological impacts are of such severity that management of the Asian clam, if detected early, is critical.

"Two reports from Tahoe find the Asian clam is an invasive species that requires intensive management in order to protect the ecological health and recreational enjoyment of the lake:

"1) In Tahoe [Asian clam] distribution is increasing and it has been linked to the large algal blooms of summer 2008. Because of its ability to alter habitats as well as to bioconcentrate calcium, it [plays] a large role in the facilitation of invasion for other bivalves such as quagga mussels. (Wittmann, et al., 2008)

"2) The rapid expansion of Asian clams in one year combined with demonstrated potential to alter the ecology of the lake via unprecedented levels of algal biomass in the nearshore represents a major new threat to Lake Tahoe. (Chandra, 2009)"

Two years later Chandra and Wittmann co-authored a paper with nine Tahoe researchers and reversed the claims they had made to the Lake George task force and concluded that Asian clams decrease algae in areas near their beds. Even though the 2013 study, *Development of a risk model to determine the expansion potential environmental impacts of Asian clams in Lake Tahoe,* makes clear that Asian clams aren't the bio-fouling pests they had been portrayed to be, the first paragraph of the 86-page report repeats the exact same wording used five years earlier in the study that justified Tahoe's Asian clam management plan:

"Because of its economic and ecologic effects, [Asian clams are] the most important non-indigenous aquatic animal in North America"—a remarkable claim considering that quagga and zebra mussels are non-indigenous. The $321,000 study, funded by the U.S. Forest Service, says Asian clams do $1 billion of damage annually in the U.S.—an exaggerated claim, even if the cost of all the university studies and the programs to prevent and control them are included in the damages, and then multiplied by ten.

The *Lake George Newsletter* reported: "We don't want to follow in the footsteps of Lake Tahoe, where action was delayed for years. Tahoe's area of infestation rapidly increased from a few acres to over 200 acres. Extensive clam beds and accompanying algal blooms now dominate what were once beautiful and clear bays. Lake Tahoe is past the point where eradication is possible." Further exaggerating the Asian Clam "epidemic" in Lake Tahoe, Village of Lake George Mayor Robert Blais said in the Public Television documentary, *Lake Defenders*, "Now there are entire [Lake Tahoe] bays closed to boat traffic."

When New York State lobbyists pushed for legislation to regulate the sale, purchase, and transport of aquatic invasive species, they used Tahoe's "Asian clam problems" as an example of what not to allow. Showing distorted pictures of the 2008 Marla Bay algal bloom, the images were captioned: "Asian clams excrete high amounts of nutrients from the sediments and make it readily available to algae—creating algal blooms and turning once clear water green." An April 2012 article in the New York *Legislative Gazette* exclaimed, "When Asian clams were found in Lake George in August 2010, researchers from Lake Tahoe visited to share what they had learned. Their message was loud and clear: Don't wait! Take action now!" Lake George became the second lake to use rubber mats to suffocate Asian clams.

A week after the *Legislative Gazette* story, an Associated Press article from Lake George said, "Scientists have launched an all-out assault on a tiny invasive clam that has clouded the cobalt blue bays of Lake Tahoe and now threatens to befoul the crystal-clear waters of a beloved Adirondack lake. This week, divers began laying 900 50-foot-long plastic mats on the sandy bottom of Lake George to smother the hundreds of thousands of Asian clams discovered last August in a 6-acre area of the lake.

"Peter Bauer, executive director of the Fund for Lake George, said: 'Lake George is in a unique position in that we detected the clams early, and we believe that this is the only site they are in.' Lake George has strong oversight from non-profit groups, scientific organizations, and regulatory agencies, who quickly mobilized a $400,000 eradication effort. 'We hope we have a chance to do here what, so far, has not been successful in other places. Lake Tahoe is a cautionary tale for us.' Bauer noted that scientists in Lake Tahoe were unable to mount an immediate eradication effort because of regulatory, financial, and other reasons. Today, Asian clams have spread to over 220 acres there."

A year later, the early optimism of eradicating Lake George's clams waned. An October 2012 *Tahoe Daily Tribune* column reported on the progress of Lake George's program: "As the Lake Tahoe Invasive Species program prepares to start laying bottom barriers [in Emerald Bay], the Lake George Park Commission on the opposite end of the country gets ready to tackle four newly discovered clam beds with the same method.

"Commission Executive Director Dave Wick said the efforts to eradicate Asian clams in the New York lake are modeled after the science and the work done in Tahoe. [Lake George has] had mixed success in treating the 35 acres of infested lake bottom [eight times the initial estimate], but the population still hasn't become completely unmanageable. It's an expensive project—

about $2 million in two years—and one that needs almost 100 percent eradication for success.

"In bids to potential donors, Wick said his sales pitch doesn't vary. There's still time to save Lake George, and organizations need to act before it becomes too late. 'We don't want to end up like Tahoe where the clams are simply impossible to manage. We can reference some of Tahoe's history and use that as a call to action.'" Two weeks after the article on Lake George's Asian clam eradication program, Geoffrey Schladow contradicted Wick and said that "complete eradication [of Asian clams] just isn't feasible [and] isn't possible anywhere."

Six months after Wick's plea to donors, the Lake George Asian Clam Rapid Response Task Force (LGACRRTF) issued a press release saying: "Due to limited funding and resources, treatment efforts in the fall of 2012 were site dependent. The new sites are all relatively small (each one less than 1.5 acres in area), so they were all treated first and in their entirety, in an attempt to achieve site specific eradication at these locations.

In the two years following their plan to contain and eradicate Asian clams in a single 4 to 6-acre site, the clams were discovered in 12 more sites covering the length of the 32-mile lake—and then the clam problem got worse. A September 2013 press release by the task force said: "Spring survey results suggested that the seven acres of mats placed on the lake floor last winter successfully killed off populations of Asian clams in Lake George. However, a two-week lake-wide survey reveals that the invasive clams are spreading beyond the treated areas."

Instead of conceding that the eradication plan had failed and that control was fruitless, the executive director of the Lake George Association said, "We have a sound and proven method to kill off the clams that we treat, but it's not enough to contain them. While it is unfortunate that we have moved beyond an eradication and containment strategy to a long-term control operation, we are still able to build upon our past successes to learn more about the clams so that we can better manage them."

Wick was not alone in his conviction that what was needed was more scientific information to better manage the clams, and the association's science advisers agreed. Meg Modley, Lake Champlain Basin Program's AIS Management Coordinator said, "The detection of the juvenile stage of the clams is the challenge; control has been successful where clams larger than 2mm are found. We need to continue surveys to document the rate of spread and study the species in more detail to understand what we can do to best manage the population."

Darrin Fresh Water Institute Director Sandra Nierzwicki-Bauer said, "This latest news demands targeted research aimed at understanding the lifecycle of the Asian clam, which will open new doors to more effective treatments that can stop the spread and limit impacts to the lake to the fullest extent possible. This is not the time to throw in the towel, but, rather, to redouble our efforts to develop better methods for treating the problem."

The task force didn't throw in the towel—or develop a better control method—and a year later all optimism evaporated when Wick announced, "Given the high cost for [clam] treatments coupled with the densities at the new sites, it appears that the best use of funding is for scientific study on reproduction and in-lake transport, which will enlighten the future management and control efforts for Asian clam."

As at Tahoe, Lake George doesn't have an Asian clam "problem." The clams cause no harm, and there is no reason to "treat" them. But echoing TRPA, the task force says the clams excrete fecal matter, which encourages algal growth and reduces water quality; crowd out native species and reduce biodiversity;

injure swimmers with sharp shells; and clog water intake systems—all grossly incorrect claims. But unlike at Tahoe where researchers abandoned the control program and investigated using molluscides, electric rakes, and bottom-scraping machines to control the harmless crustaceans, during the winter of 2015-16 at Lake George $30,000 was spent to cover 2 acres of newly discovered clams at Rogers Rock state campground with mats.

Lake George not only learned how to control Asian clams from TRPA and its science advisers, but also implemented Tahoe's AIS prevention method: boat inspections and decontaminations. In November 2011, TRPA's Ted Thayer met with members of the Lake George task and park commission to discuss Tahoe's boat inspection program. After two years of intensive lobbying, in May 2014 a mandatory boat inspection program began at Lake George.

Eric Siy, Executive Director of the Fund for Lake George said, "For something like this to work, you have to control access to the lake. We have to look at invasives as biological terrorists, and we live in a post-TSA world. … The Adirondacks are an island in a sea of invasives. The wonder of it is, even at this late date, we can still spare the Adirondacks the fate of other places with a prevention program like this."

As the first summer of mandatory inspections ended, Chris Novitsky, the Lake George water keeper, told National Public Radio that tourism at the lake "generates an estimated $1-2 billion a year in revenue, and provides a ton of seasonal jobs." But invasive species threaten the lake's pristine nature, and Novitsky said, "When invasives take over, a lot happens. They create algal blooms, weeds clog the shoreline, and old shells pile up."

After the first year of the watercraft inspection program, *PostStar.com* theorized a totally implausible justification for the program: "This past week we heard from the Lake George Park

Commission that divers scouring the bottom of Lake George found a much smaller spread of Asian clams than last year. We're not sure if anyone knows if there is a connection with boat washing, but we would not be surprised if there was."

The first-year inspection program was aided when New York State passed a law requiring boaters to take "reasonable precautions" to prevent the spread of invasive species. Although no resources or money was allocated for enforcement, warnings will be issued to first-time offenders whose boats have "visible plant and animal matter" and escalate up to $1,000 for the fourth offense. At Lake Tahoe, an offender who bypassed inspectors before launching his boat was fined $5,000 for his first offense.

In February 2015, the *TimesUnion.com* editorialized that "the Lake George boat inspection program is regarded as a big success. The battle to stop new invasive species from mucking up Lake George's pristine waters is being won. This is what is needed to keep non-native aquatic species from being introduced into the lake. They can wreak havoc on the lake's ecosystem, resulting in significant environmental, recreational, and economic damage. By one estimate the potential loss of property values and tourism dollars could be between $9 million and $48 million."

A year later in April 2016, the Park Commission board voted to make the program permanent. Saying that the inspection program was modeled after Lake Tahoe's, it was the first of its kind in the East. Executive Director Dave Wick said the estimated cost for the 2016 boating season would be $500,000 to $520,000, and that staffing would be about 55. Considering that Lake George has more inspection and washing stations than Lake Tahoe, annually inspects about 50 percent more boats, and that there is no cost to boaters, it appears that TRPA and Tahoe RCD have a lot to learn from Lake George on how to efficiently operate an inspection program and significantly reduce costs.

Besides sharing flawed science, officials from both Tahoe and Lake George present the two lakes as ecologically similar. They aren't: Lake George has 2 percent of the water volume of Lake Tahoe, and its average summer water temperature is 70 degrees (Tahoe: July surface-water temperature averages 64 degrees, below 700 feet, a year-round constant 39 degrees); its elevation is 320 feet (Tahoe: 6,225 feet), its average depth is 70 feet (Tahoe: 1,000 feet); its conductivity is above 125 microsiemens (Tahoe: less than 95 microsiemens), its average dissolved calcium is 12.1 ppm (Tahoe nearshore: 9.16 ppm); and with more than fifty non-native species, Lake George has more nonnatives than Lake Tahoe has aquatic species.

However, Lake George is a good example of how mussels can populate hospitable microregions, but if the rest of a lake's habitat is inhospitable, they don't form lasting colonies. In 1999 at the southern end of the lake, 21,000 zebra mussels were found in a volleyball court-size area. A concrete culvert was dumping high-calcium runoff into the infested area. The mussels were removed by divers and runoff mitigation was done. Over the next decade, small patches totaling less than 4,200 mussels were removed from nine locations. Since 2011 no mussels have been found.

TERC and UNR researchers frequently communicate with their Lake George counterparts and are familiar with the New York lake's patchy history with mussels. They also know that Lake Tahoe is significantly colder and that its average dissolved calcium concentration is 75 percent of Lake George's, yet Tahoe's environmental guardians maintain quagga and zebra mussels are an imminent threat. Ironically, Lake George doesn't share Tahoe's level of concern about dreissenid mussels—knowledge formed from more than 15 years of experience in having small, sink colonies of zebra mussels periodically pop up and then disappear.

Steve Urie

— 10 —

Donner Lake: A Case Study

[Mud snails' ability to] survive for up to three weeks out of water contributes to the species' ability to expand its range. ... Ideally, the experiment would have occurred over more than 14 days, but the need for immediate information to assist with ongoing management decisions as well as funding limitations necessitated the short duration of the experiments. ... In fact, results from this, as well as other studies in the region, led to the establishment of new guidelines for boat permitting and inspection at Lake Tahoe to proactively reduce the likelihood of nuisance aquatic species introductions. ~ Mud Snail Survivability Study; Alex Kolosovich, Sudeep Chandra, Laurel Saito, Clinton Davis, Lisa Atwell

Habitat, climate, predation, water chemistry, and suitable nutrients determine whether an aquatic animal can find a niche in a foreign habitat. Physiological tolerances and adaptivity from birth to reproductive maturity determine sustainable colonization. And in the Sierras, even if a micro-organism, mollusk, amphibian, or fish has adequate habitat and sufficient nutrients when lake water warms in the summer until it goes dormant in the winter, there is no guarantee that like Lahontan cutthroats it won't succumb to parasites or a disease that it has no resistance to or be preyed upon to extinction. But the primary reason why a two-week lab test for New Zealand mud snail survivability is absurdly illogical is that the snails take a year to mature—any survivability test that doesn't cover an animal's full lifecycle in a comparable habitat has no scientific value.

In the Lower Truckee River Watershed, six large mountain lakes flow into the Truckee River to mix with the only outflow from Lake Tahoe. They are Independence, Webber, Boca, Prosser, Stampede, and Donner Lake. Snowmelt from 3,000 square miles of alpine forest feeds the river as it falls 2,440 feet over its 120-mile course from Lake Tahoe to Pyramid Lake. Donner Lake is the deepest and third largest of the six lakes and is the only one that has extensive development. It is also the most heavily used for water sports and fishing. There, cabins in the Town of Truckee crowd along 80 percent of the seven-mile shoreline. Donner Memorial State Park's beaches grace the remainder.

Many who live in Truckee were surprised when they read in January 2013 that "Inspections on motorized and trailered vessels entering Donner Lake will be mandatory starting this spring as officials attempt to stay one step ahead in the battle to prevent invasive species from breaching popular water bodies in the Sierra." All boaters who lived in northern California and Nevada knew inspections were mandatory at Lake Tahoe, but wondered why now only at Donner Lake, and not the other lakes?

After boat inspections were enacted four years earlier at Lake Tahoe, those who fished and boated on Donner Lake and the nearby reservoirs believed they had dodged a bureaucratic bullet. For three years at the Donner Lake public launch site, friendly parks and rec employees collected parking and launching fees, handed out brochures with pictures of quagga-caked boat props, and offered voluntary AIS inspections. And the launch site attendants politely deferred if a boater declined an inspection; there was a winking acknowledgement between boaters and inspectors that local lakes must be immune to quagga infestation. Otherwise, inspections would be mandatory and inspectors would also be at Donner's private launch sites and the reservoir lakes. After all, hundreds of boats from Nevada's quagga-infested lakes annually launched there, and analysts hadn't found so much as a single quagga larva in Truckee area lakes.

A *Sierra Sun* article announcing the inspections said: "'The health of local waters is extremely important to the character, natural beauty and economic vitality of the town of Truckee,' said Dan Olsen, animal services and code compliance manager. A committee from the town, Truckee River Watershed Council (TRWC), and Tahoe RCD is developing the program and fees. Teresa Crimmens, a Tahoe RCD spokesperson, said 'Inspections are done to prevent the introduction of invasive species like quagga and zebra mussels, as the locust-like mollusks could damage the region's environment and hurt the economy in the form of decreased property value and millions in lost tourism.

"'During the involuntary [sic] inspections at Donner Lake over the past few years, we collected data of where boaters were coming from. What we found is many had previously launched in bodies of water that were known to be infested. According to data collected since 2010, zebra mussels or quagga mussels do not live in Donner Lake,' Crimmens said. Other less-dangerous invasives, such as Asian clams and crawfish, have made it into its waters, however."

Tahoe RCD reported that during one three-week period in 2012 "at least 204 high-risk boats launched without being decontaminated." Crimmens didn't explain why, if Donner Lake had so many exposures and if TRPA was correct in claiming that "even one larva could multiply rapidly and the resulting population could destroy the environment and economy," there were no mussels in any of the lakes. And why after years of exposure, no lake in the entire Truckee watershed remained uninfested.

A week after announcing the program, Crimmens chaired a meeting of the Truckee AIS working group. She reported that notice for mandatory inspections at Donner Lake was published by six news outlets. Commenting on the press release, David Willoughby, a senior engineer with the California Department of Water Resources (CDWR), expressed the importance of not misleading the public by using unattributed pictures from Lake Mead and implying quagga were in local waters.

Ron Penrose, an engineer with Reno's Truckee Meadows Water Authority, disagreed and said that quagga were in regional lakes. He said that in 2011 quagga DNA was found in northern Nevada's Rye Patch and Lahontan Reservoirs. But Willoughby and Jason Julienne, a California Fish and Wildlife ecologist pointed out that the reservoirs were ecologically very different from Sierra lakes and detecting quagga DNA only meant that the water bodies had been exposed and didn't mean they would become infested. Confirming the scientists' point, in the four years following 2011 no mussels were found in the northern Nevada reservoirs.

Willoughby referenced a risk assessment, *Examination of Calcium and pH as Predictors of Dreissenid Mussel Survival in the California State Water Project* that RNT Consulting did for CDWR. The analysis said that lakes with calcium concentrations as low as Donner Lake were unable to support quagga. And Julienne cited a U.S. Bureau of Reclamation report, the *Mid-Pacific Region Dreissenid Mussel Infestation Susceptibility Assessment* that backed up Willoughby's comments. The Bureau's report didn't list Donner Lake among the water bodies evaluated, but like the Whittier study classified Prosser, Boca, and Stampede in the lowest classification for mussel infestation susceptibility. (USBR, 2011) Sudeep Chandra took four calcium readings in 2012 at Donner Lake. They averaged 5.2 ppm—57 percent of Lake Tahoe's and lower than the three area reservoirs cited as "very low risk."

Crimmens argued that "calcium levels in Donner Lake are likely increasing due to the presence of Asian clams, so it would be hard to label the lake's infestation risk as low." Her theory was repudiated by Donner Lake water quality surveys—during the prior three years, dissolved calcium had slightly decreased. The *2012 Donner Lake AIS Inventory* reported, "Another interesting pattern detected during 2012 sampling is the continued decrease in calcium concentration inter-annually. If this pattern continues during continued sampling, it would appear that these lakes are becoming more resistant to the invasion of dreissenid mussels." (Caldwell & Chandra, 2012) The *2013 AIS Inventory* reported that Donner Lake's calcium concentration continued to decline.

Crimmens, a graduate biologist and one of six Tahoe RCD staff who refers to themselves as environmental scientists, should have known quagga can't establish in Donner Lake—she included links to the studies referenced by Willoughby and Julienne and to Whittier's study in her minutes. Either she didn't read the reports or refused to accept the scientific findings from RNT Consulting, the U.S. Bureau of Reclamation, and the U.S. EPA. Another report she listed, the U.S. Geological Survey's *Pacific Northwest Aquatic Invasive Species Profile,* says that to maintain shell growth in water with low pH, quagga mussels require 25 ppm of dissolved calcium—five times Donner Lake's calcium level. (Richter, 2008)

Shown the scientific evidence, Truckee town staff delayed asking for approval to implement the boat inspection program until Sudeep Chandra was able to respond to allegations that his science and justifications for the program were flawed. A public meeting was held in late April 2013, only days before the inspections were to begin. Referring to the study that he and Kumud Acharya had begun for the U.S. Forest Service two years earlier, Chandra told the attendees that he had started a second survivability study using a broader spectrum of Lake Tahoe water instead of only water from the Tahoe Keys, and that his assessments would be available in the fall.

He told the Truckee AIS Working Group that he believed his study "may indicate that mussels can survive and potentially reproduce at calcium levels significantly lower than previously believed, [and] that we should not assume an infestation in Donner Lake is not possible." In fact, he had already completed his first attempt at his second survivability study, and it had failed to show that quagga could survive and reproduce in Tahoe Keys water, much less in water with calcium as low as Donner Lake's.

Two weeks later, the question of implementing mandatory boat inspections at Donner Lake was presented to the town council. Town staff had recommended that the proposed program be suspended for a year in order to refocus on all threats to the Truckee River Watershed, and specifically to review the science on quagga and zebra mussels. The town council voted to accept staff's recommendation, and told them to document the "multi-species AIS concern in the Truckee River Watershed"—it appeared that Truckee had put in place a balanced process to assure the efficacy of their AIS program.

At the Truckee presentation, Chandra said that "if boating inspections are based only on preventing quagga infestations, we are in deep trouble." And Crimmens backed him up saying, "The intent of the watercraft inspection program is to prevent introduction of not only mussels, but also other AIS of concern and those that have not yet been identified." It's hard to research unidentified species, but the agency's list of what their inspectors look for is "quagga and zebra mussels, New Zealand mud snails, spiny water flea, didymo (rock snot), hydrilla, and other highly invasive plants."

Behind quagga and zebra mussels, New Zealand mud snails top Tahoe RCD's list of potentially invasive aquatic species.

A month after the Truckee town council suspended mandatory boat inspections at Donner Lake, the news that three New Zealand mud snails were found in the Truckee River in Reno created regional headlines. The tiny snails should have been of no concern to Tahoe-Truckee ecologists and water managers—mud snails prefer warm, high-nutrient water over clear, cold water.

Because New Zealand mud snails do no significant harm, there has been very little research to determine threshold levels for their survival, growth, and reproduction. One threshold that has been established that is important in determining the ability of the resilient snail to colonize in Sierra lakes and streams is conductivity—New Zealand mud snails don't prosper in water with conductivity less than 200 microsiemens (μS/cm)[8], a level that is double that of High Sierra lakes and the California stretch of the Lower Truckee River.

Tahoe RCD maintains mud snails would be harmful if they established in Lake Tahoe, but like quagga, the tiny snails can't survive in the lake. However if they could, because they eat detritus, they too would improve the lakes' water quality. And an Australian study showed that the snails positively influenced the colonization of macroinvertebrates. (Schreiber, 2002) Tahoe RCD says rafters, kayakers, or anglers are to blame for spreading them to the Truckee River in Reno. Many wondered if that were true, how would boat inspections keep them out of Donner Lake?

Behind New Zealand mud snails on the list of potential Sierra AIS is spiny water flea. Not much is known about the mini crustaceans that are similar to mysid shrimp, but they are terrific food for large trout. It's still unclear if spiny water fleas can survive in the Sierra, but any organism that can establish in mountain lakes and trout feed upon is considered by most to be a

---

[8] The measurement of electrical conductance in a solution

positive introduction. Some sounded a cry of alarm when spiny water fleas were found in Lake Champlain in 2014, but biologists from Vermont Fish and Wildlife quickly calmed fears when they pointed out their benefits to the fishery and said they would have a negligible effect on the lake's native animals.

Peering into the depths of Lake Tahoe, the translucent water seems devoid of life, and most don't suspect that the water they look through contains more than 100 and possibly as many as 300 algal species. The six groups of Tahoe's algae range from invisible unicellular organisms to complex multicellular forms, such as that which formed the filamentous bloom in Marla Bay in 2008. Algal growth responds to changing water temperature, sunlight, and nutrients. In winter there is almost no algal biomass in Lake Tahoe and algae lies dormant, but as days lengthen and water warms, some algae grows faster than dandelions. Since surveys began in the 1980's, Lake Tahoe's total algal volume has been relatively uniform and closely tracks seasonal conditions.

Didymo, popularly named rock snot, is an algal species that possibly lies dormant in many High Sierra lakes. It is in most northern U.S. lakes and streams and usually lives unobtrusively alongside other algae. But when conditions are ideal, it explodes and forms thick mats on stream beds. Uncharacteristically for almost all algae, it has been found in some of the nation's clearest waterways, including oligotrophic streams. In California, didymo blooms have occurred at low elevations in the American and Yuba Rivers. However because the microscopic alga can spread in a single water droplet or on a wet swimsuit or towel, it is impossible for boat inspectors to detect.

Biologists were puzzled about how rock snot was able to form dense mats where exceptionally high water quality made any algal growth improbable. But a National Science Foundation study found that unlike other algae species, when water warms and bacterial activity increases, didymo concentrates phosphorus. The study uncovered a mechanism by which microbes synthesize phosphorus, creating a microenvironment and explaining how rock snot is able to bloom in low-nutrient streams and rivers. (Science Daily, 2011) Didymo blooms haven't occurred in the Sierras, and year-round cold water may keep the mountains forever free of rock snot.

A lot has been written about hydrilla. It was first reported in California in 1976, and by 1990 had infested many southern California lakes and reservoirs. Because the plant doesn't tolerate freezing temperatures, lasts only a day or two in dry conditions, requires nutrient-rich substrate, and seldom grows at depths of more than ten feet, Sierra lakes don't provide a suitable habitat for the aquatic plant. (CDFA, 2014) And even if it were to find a place to take root during the late summer, the plant wouldn't make it through a Sierra Nevada winter.

When asked to identify other non-native species that may threaten Sierra lakes, Darcie Goodman Collins, the Director of the League to Save Lake Tahoe, suggested golden mussels, which she described as "quagga on steroids." She pointed out that they can survive where calcium is very low. But the South American natives don't prosper in cold water. Temperature is the primary factor in initiating their gamete release, and water temperatures that dip below 46°F for more than a month are lethal to golden mussels. They have never been found in North America, but if they do make it to U.S. shores, January and February will eliminate all Sierra Nevada water bodies as a potential home. (USACE, 2013)

The LTSLT is Tahoe's most influential environmental organization. For decades the League has had a contentious relationship with TRPA and often kept the agency honest by litigating environmental matters, but LTSLT totally subscribes to

TRPA's vilification of non-native aquatic animals. In a May 2014 *Tahoe Daily Tribune* op-ed, Collins wrote: "Boat inspections are again open for business. Inspectors are true heroes in the effort to Keep Tahoe Blue. They decontaminate thousands of boats each year to ward off new invasive species. The top threats are quagga and zebra mussels, pests that multiply rapidly and have destroyed ecosystems in Nevada's own Lake Mead. It only takes one contaminated boat to forever alter the Tahoe we know and love.

"Imagine looking into Tahoe's waters and instead of a white sand bottom [one has to wonder if Goodman Collins has ever looked into Lake Tahoe—there is no white sand bottom], seeing a thick mat of mussels. Mussels concentrate nutrients, starve native species of food, and grow easily on hard surfaces like pipes, clogging equipment and causing severe economic problems. Boat inspections are expensive ... but if invasive mussels establish here, addressing the problems they cause would be far more expensive. We can't afford the risk."

Two nonnatives, Asian clams and crayfish, have been in Donner Lake for as long as most residents can remember. Besides amusing kids who put out hotdog-baited traps to catch the baby lobsters, anglers who land the lake's huge mackinaw and browns note that crayfish are a staple of the fishes' diet. And even though they say the ubiquitous crayfish are "invasive," unlike at Lake Tahoe, Tahoe RCD doesn't try to convince Donner Lake's residents that crawdads should be controlled and Asian clams should be suffocated with rubber mats. Because they have been in the lake for at least 40 years and have caused no harm, no one at Donner Lake even considers them to be invasive.

Tahoe RCD's claim that boat inspections are an effective way to prevent Asian clams from spreading poses a dilemma for the agency: If the reservoir lakes, which have no clams, are to be equally protected, shouldn't Truckee anglers who fish at Donner

Lake be required to decontaminate their boats before launching at Boca, Prosser, or Stampede Lakes where there are no clams? Tahoe RCD addressed the issue during 2014 and assigned an inspector to monitor the three lakes, which are spread over a 600-square-mile area. In 2016 the inspector was eliminated, and at Boca and Stampede self-inspections were implemented.

As it is increasingly evident that each of the non-native aquatic species Tahoe RCD suggests might invade the High Sierra can't survive there, AIS threats become more generalized, and agencies and organizations increasingly use the threat of "species not yet identified" to justify inspections. Goodman Collins stated the attitude of many environmental watchdogs when she wrote that the League is "charged with advocating for strong environmental regulations. As such, it is our responsibility to err on the side of caution and support an effective and minimally intrusive inspection program that could protect the lake from any harmful, known or unknown, invasive species infestation. The beauty of boat inspections is that it helps to reduce the threat from a number of species, including, but not limited to, quagga."

Most people aren't threatened by unknown species, and virtually all boaters don't believe that paying up to $150 for a needless inspection when they launch in Lake Tahoe is minimally intrusive. But TRPA has so thoroughly indoctrinated the boating community about Tahoe's AIS perils that most believe that it's not only the law but also their environmental duty to assure that they aren't the one who causes a devastating infestation. But most don't believe that suspending judgment on environmental impact, assessing burden on users, and evaluating program efficacy should be subjugated to protecting Lake Tahoe from species not yet identified.

Seemingly with the number of agencies and organizations invested in preventing AIS from infesting Lake Tahoe, there would be a large amount of scientific data on the ability of nonnatives to invade and harm the lake. After the Truckee town council postponed inspections for a year to evaluate whether Donner Lake was susceptible to any AIS, Crimmens and Tahoe-Truckee agency managers were asked for data or studies that supported the notion that any of Tahoe RCD's list of potential aquatic invaders could establish in Tahoe-Truckee area lakes and to suggest other species that may be able to infest them.

No supporting research or data was provided and no other potential aquatic invasives were identified. Considering that there are 6,000 freshwater mollusk and 14,000 freshwater fish species, it's remarkable that Tahoe's ecologists and aquatic biologists can only come up with three mollusks, two plants, spiny water flea, and rock snot as threats to Sierra lakes, and that they can provide no documentation to support their claims that those potential invaders are able to establish in the lakes.

Paying for environmental programs is always a struggle. The press release announcing mandatory inspections at Donner Lake said inspection fees would completely offset the program's costs. In fact, fees—projected to be about $40,000—would only cover about half of Crimmens' annual salary. In 2012, Tahoe RCD received $179,455—60 percent of the voluntary program's total cost of $303,655—from the Truckee River Fund, a water quality program funded by the Truckee Meadows Water Authority.

Ron Penrose, the manager of the Truckee AIS program for the TMWA, justified his Reno agency underwriting the inspection program saying that TMWA didn't want to incur the "tens of millions of dollars of cost" that their Las Vegas counterpart, the Southern Nevada Water Authority, was spending to control quagga. That sounded fiscally prudent, but he was wrong about quagga's ability to establish in Donner Lake and about the cost of controlling Las Vegas' quagga infestation. Stating mitigation costs for the SNWA similar to those cited in the *Quagga Mussel Risk Assessment*, in 2013 Penrose wrote in the *Reno Gazette-Journal* that SNWA's quagga mussel control program had cost $38 million—an amount that is greater than the annual total spent in North America on dreissenid mussel mitigation. SNWA's total cost for quagga control from when mussels were found in Lake Mead through the program's first six years was less than $1 million.

Taking its water from Lake Mead, the SNWA supplies water to 2 million desert residents. Lake Mead's calcium is greater than 90 ppm, and there mussels grow larger and reproduce twice as often as those in most lakes where quagga can establish. It was initially projected that SNWA's annual costs for mussel control would be $1 to 4 million. (Flanzraich, 2008) The projected costs didn't materialize, and SNWA's Peggy Roefer reports that her agency's costs for equipment were minimal. They had to buy one $10,000 chlorine pump, and annual treatment and chemical costs are about $150,000. Ironically, in 2012 the Las Vegas water company paid less to abate the nation's most serious quagga infestation than Penrose's Reno water company paid to subsidize Tahoe RCD to protect Donner Lake from mussels. Stated in other terms, each of TMWA's 93,000 customers gave about 2 dollars a year to the Tahoe RCD to prevent quagga mussels from infesting Donner Lake while Las Vegans pay 8 cents per resident to fully mitigate impacts from the nation's worst mussel infestation.

Three months after the Truckee Town Council was assured by Tahoe RCD that not only would inspection fees for Donner Lake boaters be minimal but that the town would make some money on the program, the *Reno Gazette-Journal* reported, "A pot of federal money that's been key to funding boat inspections [at Tahoe] is now drying up. This summer officials raised the fee to decontaminate vessels, closed one inspection station on the lake's west shore, and will reduce hours. But this year's issues are nothing compared to what could be coming next year, when the boat inspection effort faces a shortfall of up to $750,000."

The doubling of inspection fees was postponed for a year when the Bureau of Land Management pledged to underwrite the 2015 program—spreading the cost of Lake Tahoe's needless boat inspections to all of the nation's taxpayers. A week later, saying "a strict inspection program has managed to prevent the worst of the invasive species, especially the quagga mussel, from finding its way into [Lake Tahoe]," a Forest Service representative began the appeal for funding beyond 2015. In early 2015, TRPA proposed that California and Nevada split the annual $750,000 shortfall for boat inspections, and two months later the states agreed to equally fund half the annual cost of the federal program.

After his Truckee presentation in April 2013, Sudeep Chandra wrote to town staff that the results from the 2008 *Quagga Mussel Risk Assessment* had surprised him. What he failed to tell the town, or any of the agencies that rely on his work, was that his surprise had been indeed unjustified—only a few weeks earlier in his follow-up survivability test, 19 of 20 quagga died within 20 days. And he wrote to the Town of Truckee: "In regards to whether you misinterpreted what I said, and if I was opposed to a risk assessment, I think your suggestion [to conduct] a local risk assessment for a variety of invasive species that may pose a threat to our region is an excellent idea.

"I believe this is the position I stated in the meeting, and if I didn't it should be made clear in this email. It is very important for communities to make a self-assessment for the risk that invasive species may pose to their water bodies and the uncertainty posed in such an analysis. Often we make decisions in a world of uncertainty. The good news is that we can alter those policies as new scientific information emerges."

Although Chandra redid his experiments, very probably if he had told Truckee officials or the working group that only one mussel out of 20 had survived for more than 20 days in his 90-day experiment, the council would have delayed implementing mandatory boat inspections for a second year while he performed more tests. When Teresa Crimmens was asked about Chandra's preliminary findings, she said he hadn't replied to her inquiry.

Crimmens took charge of Truckee's AIS risk assessment process, rejected all science that conflicted with her opinions, and three months after the town council instructed its staff to refocus "the discussion to include all possible aquatic invasive species threats to the Truckee River Watershed," Crimmens, her manager, and two staff from the TRWC submitted the *Truckee Regional AIS Prevention Program (TRAISPP), AIS Vulnerability Assessment"* to the town and invited Donner Lake stakeholders to a presentation of their assessment.

The sparse 13-page document concludes: "Water quality data in the Truckee Region does not preclude the establishment of species examined here and indicates high suitability for some species." It cites no water quality data and provides no biological analysis or habitat requirements for the examined species: quagga and zebra mussels, New Zealand mud snails, spiny water flea, rock snot, watermilfoil, curlyleaf pondweed, and hydrilla. But in fact, Donner Lake's water quality does preclude each of the animals listed, and climate and substrate exclude the aquatic plants.

Focusing on mussels, to buttress their recommendation six quagga studies are referenced in the *AIS Vulnerability Assessment*. Remarkably, data from five of the studies verifies that mussels can't establish in Sierra lakes. The sixth was a history of mussel research and presented no conclusions. And conspicuously, Chandra's *Quagga Mussel Risk Assessment*, the only study that claims quagga mussels may be able to propagate in low-calcium water wasn't included.

Discussion at the Truckee presentation of the vulnerability assessment was contentious. After heated exchanges over the measures needed to protect Donner Lake from quagga, David Willoughby, a California Department of Water Resources engineer whose job includes collecting water quality data and performing AIS risk analyses, calmly explained that unlike Asian clams that don't form shells during their veliger stage, quagga do, and that is one of the reasons why quagga need higher water calcium levels than Asian clams. And he concluded by saying that quagga can't survive long-term in Truckee area lakes. Agreeing to expand the scope of the assessment, Crimmens adjourned the meeting. It was the last public input into Truckee's *AIS Vulnerability Assessment*.

Habitat and water quality are the most important factors in determining species establishment. The Truckee assessment doesn't mention habitat, and although Tahoe regional lakes' water quality is among the world's highest, it rates Donner Lake's water quality as "Medium." Nationally, Lake Tahoe trails only Crater Lake in clarity, and arguably, Donner Lake has higher water quality than Tahoe's, and its nearshore dissolved calcium is less.

The Truckee vulnerability assessment states: "Water quality data is inconclusive on mussel survivability, but water quality is not limiting New Zealand mud snail, Asian clam, Eurasian watermilfoil, curlyleaf pondweed, hydrilla, or rock snot." Water quality data is not inconclusive on mussel survivability—all research but Chandra's says quagga can't propagate in low calcium water. Exceptionally low conductivity excludes New Zealand mud snails from High Sierra lakes. Asian clams are harmless, and freezing temperatures and annually lowering reservoirs 10 to 30 feet prevent aquatic plant infestations.

At the presentation of the vulnerability study, Lisa Holan, a UNR aquatic ecologist and board member of the Mountain Area Preservation Foundation, said she didn't believe quagga could infest Donner Lake, but at the town council meeting, she spoke in favor of boat inspections, saying "To just dismiss the experiences of so many other communities across North America as irrelevant because they're not our specific lakes, I think is very dangerous. Every lake is unique, and [dismissing the possibility of invasion] didn't protect those lakes." It is correct to dismiss data and ignore studies on lakes that are dissimilar to High Sierra lakes—lakes are indeed unique, and that is why water quality analyses and risk assessments are performed.

Aquatic animals are as biologically different and habitat sensitive as their terrestrial relatives. And because habitats differ greatly, it's critical to analyze water chemistry, protective cover, climate cycles, and nutrient availability, to determine if potentially harmful nonnatives can survive in specific ecosystems. And even if establishment is possible, it doesn't mean damaging infestations will occur. Zebra mussels are an expensive problem in Lake Heron where calcium reaches 30 ppm, but in Lake Superior, North America's largest lake and Lake Heron's northern Great Lakes neighbor, near-shore calcium only approaches 16 ppm, and zebra mussel colonization is sporadic but never harmful.

Whereas economic damage caused by nonnatives is easy to calculate, environmental harm is difficult to assess. Whenever a new species establishes, the food web is altered and some natives may be out-competed. But that doesn't mean the natives will be driven to extinction or even adversely affected. Introducing non-native trout for sport fishing is an example of how disrupting natural ecosystems isn't necessarily bad. Fishing writer Bruce Ajari believes that nonnatives threaten Donner Lake's minnows. He's correct—big fish prey on little fish. Huge trout are pulled from the lake, and besides eating crayfish, they eat minnows. Unarguably, the introduction of mackinaw, brown, and rainbow trout altered the lake's food web, but its native minnows aren't endangered.

TERC Assistant Director John Reuter feared Asian clams may consume so much plankton that native fish would starve, but the clams have quietly lived in Donner Lake for 40 years, and browns and mackinaw are thriving at historically high populations. Tahoe RCD says Asian clams and crayfish are invasive to the area's lakes, but neither cause any harm. As significantly, by filtering algae Asian clams improve water clarity, and because they are scavengers and feed on detritus, crayfish improve water quality.

Crimmens represented to Truckee town staff that the AIS vulnerability study was the consensus findings of the AIS working group, and in January 2014 the Town of Truckee became the nation's first municipality to enact an AIS ordinance, and Donner Lake gained the distinction of the lake with the nation's lowest calcium concentration that has mandatory boat inspections. At summer's end, the Truckee River Watershed Council posted on their website: "Donner Lake has finally gotten the recognition and support that it deserves. Beginning in 2014, watercraft inspections are now mandatory before launching into the lake. The Truckee River watershed stakeholders saw the need to initiate a community-wide approach to stop the spread of aquatic invasive species, such as the quagga mussel, New Zealand mud snail and Eurasian watermilfoil. ... The introduction and establishment of invasive species can alter natural ecosystem function, displace native species, increase pollutants such as algal blooms, damage boat propellers, and cause economic losses for the community."

All are false or greatly over-stated claims that are nearly impossible in the High Sierra, but the greatest misrepresentation was that local stakeholders saw a need to initiate a community-wide approach—only those who benefited from the inspection program supported it. With funding from a Reno water company, Tahoe RCD simply expanded their Tahoe program to Truckee. By ignoring scientific evidence and misrepresenting studies, two non-technical staff members from the Truckee River Watershed Council and two Tahoe RCD managers seized the process, represented themselves as experts, and recommended mandatory boat inspections at Donner Lake.

As can be expected when science and objectivity are pushed aside, the ordinance is riddled with factual errors. It lists several irrelevancies and states an incredulous endangerment: AIS "pose a significant and imminent threat to Truckee's sewer collection and treatment infrastructure"—an impossibility first advanced by Chandra and later repeated by Crimmens that has to make the town's waste treatment plant engineers cringe.

The Truckee ordinance highlights some of the issues that confront AIS regulation: regional inconsistencies, enforceability issues, and bias from those who benefit from prevention and control programs. California ecosystems are exceptionally diverse, and creating state-wide AIS regulations is impractical. Whittier's mapping of the United States for the risk of mussel infestation illustrates how ecosystems don't follow jurisdictional boundaries—southern California is in a region at highest risk for infestation; risk along the north coast is "highly variable;" and the Sierra Nevada is in the lowest risk classification.

Truckee rejected the federal ANS Task Force AIS definition and wrote their own: "Aquatic Invasive Species are aquatic animals and plants that have been introduced into waterways in which they do not live naturally. They have harmful effects on the natural resources in these ecosystems and the human uses of these resources"—a sufficiently vague definition that allows the classification of all aquatic nonnatives as invasive. The ordinance ignores the Truckee River, which flows for seven miles through the sprawling town, and the nearby reservoir lakes are under federal jurisdiction and aren't subject to the town ordinance. However because no non-native species threatens the river or the reservoir lakes, they too will remain AIS free, and the Truckee AIS plan will be lauded as being highly effective.

TRPA boasts that they have the nation's most aggressive AIS prevention program, and nowhere is the politicization and hype of AIS programs greater than in the Tahoe-Truckee region. Aquatic invasive species regulations are only beneficial if the AIS-related harm they prevent or control outweighs the cost of administering and enforcing the regulation. TRPA and Tahoe RCD consistently say that AIS prevention is less expensive than control, but when their managers are asked to give an example where that is true, they are unable to.

As at Tahoe, in Truckee no AIS-related harm has occurred or has a species been identified that can establish and harm area lakes. Nevertheless, the *AIS Vulnerability Assessment* said Donner Lake was at "High Risk" for AIS infestation. That is incorrect, and an emotional campaign to implement mandatory boat inspections used erroneous data, promoted pseudoscience, created false cost expectations, and misrepresented results from scientific studies to foment unfounded fears of ecological and economic harm at a California lake to launch an unnecessary $300,000 per year boat inspection program that was underwritten by Nevada water utility customers.

— 11 —

Financial Delusions

The prospect of domination of the nation's scholars by federal employment, project allocations, and the power of money is ever present—and is gravely to be regarded.
~ Dwight Eisenhower, Farewell Speech

It has been observed that Tahoe's AIS problem may be that it has too much money to solve too few problems. The 2009 *Lake Tahoe Region AIS Management Plan* states: "The Lake Tahoe Region has secured $5.2 million dollars to spend on AIS from 2007 through 2009. While a significant sum, professional experience suggests that this level will need to be sustained, if not increased, as the AIS problem at Lake Tahoe matures."

Five million dollars is a significant sum to wrangle from taxpayers. (More than $8 million was actually spent on AIS over those three years.) It's about the same as is spent annually in the Tahoe Basin on police operations, and it's more than the State of Michigan, which has 3,200 miles of Great Lakes shoreline, 11,000 inland lakes, and hundreds of zebra and quagga mussel infestations spent on AIS prevention and control over the same period. In 2007 there was no AIS problem at Tahoe to mature. And unless 35 tons of rubber matting in Emerald Bay is considered an AIS problem, there hasn't been one. But that didn't prevent

TRPA from creating AIS threats and organizing a task force to project what havoc would be unleashed if one were to exist.

There are many legitimate demands on always-limited public resources, and when government agencies compete for tax dollars, appropriately programs that provide the greatest taxpayer return get the biggest slices of the budgetary pie. Funded by the California Tahoe Conservancy and the Army Corps of Engineers and coordinated by TRPA's Ted Thayer, representatives from 11 federal and state agencies signed off on a Tahoe AIS management plan that contains an impressive display of graphs, pie charts, and tables and concludes that the "combined economic impacts to recreation value, tourism spending, property values, and boat/pier maintenance, when evaluated over a 50 year period, is estimated at $417.5 million (present value), with an average annual equivalent value of $22.4 million per year." The plan devoted 30 of its 256 pages to detailing the potential economic impacts from AIS, which not only underscores the enormous effort but also the elaborate planning and creativity needed to justify the prevention of non-threatening, non-native species.

Few places offer greater outdoor recreation opportunities than Tahoe, and the management plan says: "Approximately 2.9 million people participated in [Tahoe] beach activities during the 2008 summer, [and] it is reasonable to expect a 10–20% decrease in participation in beach activities as a result of AIS infestation [caused by] decaying organisms releasing foul odors and attracting insects, hazardous invertebrate shells washing onto beaches, and adverse effects to water clarity.

"Approximately 2.4 million [Tahoe] visitors participate in swimming during the summer season. However, the shallow nature of the swimming beaches means that AIS can drastically affect swimmers. Plants are the main concern for swimmers.

Dense, vine-like plants like Eurasian watermilfoil is not just annoying to swimmers, they are hazardous. A swimmer can become entangled in milfoil, possibly leading to drowning. Dense mats of AIS growth may be left exposed and will decay, likely emitting noxious odors and will generally be offensive to swimmers. It is reasonable to expect a 20–80% decrease in swimming, depending on the density of vegetative growth.

"Power boating participation may be reduced 10–30% depending on the extent of aquatic vegetation in shallow areas. This estimation is based on the fact that AIS are unlikely to directly impact boating activities in the middle of the lake; however, getting from the shore to the middle will be challenging, particularly in areas with dense vegetation. Power boats will likely be able to continue operation, but the lake will begin to seem more crowded as all boaters must move further from shore to avoid the invasive vegetation that can harm propellers and make the water unfavorable to skiers.

"Canoeing and kayaking are likely to be most impacted by weed infestations because these activities are concentrated in the nearshore environments where the lake bathymetry is conducive to infestation, making paddling difficult. Weeds and other AIS can also impact water clarity, especially in nearshore areas. Because of this, a reduction between 20–40% is reasonably expected.

"A survey indicated that 20% of the visitors to Lake Tahoe participated in fishing, amounting to around 914,000 fishing visits in 2008. Most anglers are likely to fish in the shallow areas of the lake that are more susceptible to AIS infestation. For anglers, invasive weeds can be aggravating and possibly damaging to a boat. Fishing from shore is not desirable when casting into a dense mat of aquatic vegetation such as Eurasian watermilfoil. AIS not only reduce the availability of catchable fish, they also reduce anglers' access to them. The explicit impacts of warm-water fish at Lake Tahoe are not fully known; however, all invasive fish compete with native and sport fish for food resources. The presence of invasive fish has the potential to damage food webs and disrupt ecosystem function. Sport fishing in the Great Lakes was reduced from 10 to 35% as the result of AIS." Although having remarkably dissimilar ecosystems, the report's authors reduced sport fishing at Tahoe due to invasive species 10 to 35 percent.

After using a complex logic model to calculate reductions to recreation value from AIS, the management plan states: An "estimated 2% visitation reduction yields recreation loss of approximately $1.3 million, while a median estimate of 5% yields $3.2 million, and 10% yields $6.4 million in lost value." Showing restraint, the task force took the most conservative number, inflated it annually by 4.875 percent over 50 years, and projected the loss each year in recreation value due to AIS to be $1,751,000.

As statisticians and economists would point out, allocating present value to an unquantifiable event that might happen is poor mathematics—and terrible logic. Through 2014 there was no economic loss due to AIS at Lake Tahoe. Though TRPA would probably say that boat inspections have saved taxpayers and homeowners at least $134 million, some economists would counter that Tahoe's AIS management plan's economic loss projections were obviously off by $134 million and will continue to be overstated at the rate of $22.4 million a year.

Economic losses from decreased recreational activity would, of course, compound due to a loss in tourism. The plan summarizes: "An AIS infestation could cause significant decreases in recreation participation and losses in recreation value. These decreases have direct effects on the tourism revenue Tahoe receives. When considering the impacts on the local economy's revenue, it is apparent that even a small reduction in visitation yields large losses in revenue. A decrease of 2% in visitation would result in about $6.1 million less dollars entering into the local economy." AIS infestations were projected to cause tourism spending to annually suffer an $8,412,000 loss.

Owners of Tahoe's multimillion-dollar lakefront homes aren't affected by tourism spending, but they are affected by a loss in property value. The AIS management plan states: "The assessed values of lake front parcels are sensitive to the quality of lake access. For example, a pier is a high value-added feature, as is beach. However, the benefits of either of these features might be diminished by the presence of AIS. Invasive plants can make the property's lakefront un-swimmable by entangling the legs of swimmers. Both plants and invertebrates wash up on shore when they die, leaving a foul smelling beach full of sharp mussel shells. In addition, plants can ruin a pier's functionality by making access to it difficult without the use of a weedless propeller. Invasive plants can also destroy habitat for native species of fish and flora while fostering mosquito reproduction, making properties less attractive to anglers and beach visitors.

"Aesthetics are also important in the valuation of a property. A home's value may decrease as a result of diminished aesthetics like dirty beaches and reduced water clarity. Given the importance of clarity at Lake Tahoe, loss in clarity is likely to have an effect on a property's value. ... [But at Lake Tahoe] it might not affect property values as quickly as by the effects of invasive nearshore plants." After crunching the numbers, it was calculated that the annual loss to property values would be $8,728,000.

Dreissenid mussels in Lake Tahoe might actually increase lakefront property values. After two years of researching the relationship between zebra mussels and property values, Martin Meder, a University of Wisconsin-Oshkosh student, presented a paper that showed real estate on lakes in 17 Wisconsin counties rose 10 percent on lakes that have zebra mussels over those that don't. The property value increase was attributed to algae-munching mussels greatly improving clarity. (Bergquist, 2014)

The management report says, "The main concern that AIS present with regard to water supply is that quagga and zebra mussels biofoul freshwater intake pipes. This invasion requires costly maintenance or periodic replacement of pipes and can result in the loss of filtration exemption due to the presence of mussels and plants in the water intake systems that raise human health concerns." Considering that quagga can't propagate in the lake, and it costs the Las Vegas water company $150,000 a year for issues related to quagga mussels, the annual estimate of $2,001,000 to Tahoe water companies appears high.

The AIS management report's Potential Economic Impact section cites the *Detroit River and Lake Erie Basin Indicator Project,* but it doesn't cite *A calcium-based risk assessment for zebra and quagga mussels,* which is the basis for all of the nation's—except Tahoe's—dreissenid mussel risk evaluations. But that didn't deter the report's authors from warning that boats in Lake Tahoe "may need to run their engines every few days to prevent mussels from colonizing the cooling system. Winter dry storage will be mandatory to avoid mussel damage, or if boats are permanently stored on the lake they may need a coat of biocide bottom paint to keep mussels from growing on the hull.

"AIS also impact boaters and business owners who have private piers and docks. Piers may also degrade and wear more quickly as a result of mussel biofouling penetrating the piles. Floating docks and buoys can be weighed down by mussel colonies and will require periodic cleaning, replacement, or reinforcement to remain functional." The projected annual cost for boat and pier maintenance due to mussels: $1,536,000.

Add up all of the annual economic damages AIS would cause if they weren't prevented by boat inspections and it is conservatively $22.4 million. It's difficult to believe that the 71 scientists, consultants, academics, and agency managers who lent their names to the construction of the *Lake Tahoe Region AIS Management Plan* were analyzing Lake Tahoe and not a Florida swamp or Louisiana bayou. But as Geoffrey Schladow told the 2011 Lake Tahoe Aquatic Invasive Species Forum, while displaying a slide of supporting members' logos, "This is the game we play. Many of you are from agencies. If your logo's not here, then it's [someone's] fault. We have member support from everybody."

It's also hard to believe that none of the technically-trained people who contributed to the management report didn't object to the hyperbolic and totally speculative plan to combat what might happen if quagga mussels and non-native plants infested Lake Tahoe. But as Jerry Franklin, a professor of ecosystem analysis at the University of Washington explains, "A lot of environmental messages are simply not accurate, but that's the way we sell messages in this society. We use hype. And we use pieces of information that sustain our position." The authors of Tahoe's AIS management plan not only did that, but also fabricated numbers to justify their budget requests.

TRPA acts as fiscal agent and is tasked with oversight for Tahoe's AIS management plan. The plan requires updating every five years, and when the plan was released in 2009, it reflected an understandable over-abundance of caution. However, in the five years after plan adoption, there were no new introductions or AIS harm at Tahoe, and as importantly from a scientific perspective, there was no harm to any Sierra lake that didn't have inspections. Tahoe RCD, TRPA, and their researchers were given most of the data presented in this book, and it was hoped that the 2014 plan update would reflect the current data and the positive regional experience. But the five-year update is mostly a restatement of the 2009 plan. One notable change is that the threat of potential annual costs due to aquatic invasive species was more than tripled from the "conservative" approximation of $22.4 million to $78 million per year. Calling the plan "a model for the nation," it was approved by the federal ANS Task Force in August 2014.

In *The Costs of Aquatic Invasive Species to Great Lakes States,* the Anderson Economic Group says, "While we cannot provide a single number for the total cost imposed by AIS, it is likely that the overall aggregate level of cost to the Great Lakes region is over $100 million annually." (Rosaen, 2012) Lake Tahoe is susceptible to none of the Great Lakes aquatic invasions, yet TRPA says that boat inspections are preventing potential annual economic damage due to AIS in Lake Tahoe at a cost nearly equaling that which the entire Great Lakes region incurs. The Anderson Group's report is the most authoritative cost analysis on the total AIS impact to the world's largest freshwater ecosystem, and although $100,000 million is a lot of money—equal to TRPA's annual Environmental Improvement Program budget—it's well below that quoted by government agencies.

The most authoritative cost analysis on the impact of dreissenidae is a 2006 study by Cornell University. It reports that "from 1989 to 2004, the total economic cost for dreissenid fouling in all electric generation and water treatment facilities in North America was estimated to be approximately $267 million," an annual average of $18 million, or less than what TRPA projects the cost of a quagga invasion at Tahoe would be (Connelly, 2007).

Prior to the Cornell and Anderson studies, a 2005 National Center for Environmental Economics (NCEE) paper, *The Economic Impacts of Aquatic Invasive Species: A Review of the Literature* was the primary source for AIS cost data. Unlike the Anderson study that considered 19 government and academic studies, but used only "eight that had credible estimates," the NCEE review didn't evaluate its data for reasonableness or accuracy, and many of its statistics are from the U.S Office of Technology Assessment, which closed in 1995, and are no longer verifiable.

Government agencies aren't the only source for highly exaggerated cost data and questionable science. U.S. Fish and Wildlife financially supports the University of Auckland's Global Invasive Species Database, where it's reported: "Introduction of Asian clams into the United States has resulted in the clogging of water intake pipes, affecting power, water, and other industry. Nuclear service water systems (for fire protection) are very vulnerable, jeopardizing fire protection. [Nuclear power plant water intake is for cooling.] In 1980, the costs of correcting this problem were $1 billion. [This appears to be the source Tahoe researchers used in the *Development of a risk model to determine the expansion potential environmental impacts of Asian clams in Lake Tahoe* for the annual cost in the U.S. due to Asian clams.]

"Asian clams cause problems because juveniles are weak-swimmers, and consequently they are pushed to the bottom of the water column where intake pipes are usually placed. They are pulled inside the intakes, where they attach, breed, and die. The intake pipe becomes clogged with live clams, empty shells, and dead body tissues. Buoyant, dead clams can also clog intake screens." (Shyama, 2005) Even if this researcher confused clams with mussels and billions with millions, his science is wrong, his mechanics are absurd, and his costs are irresponsibly overstated.

Because it provides convenient cost justification for AIS studies and programs, the decade-old NCEE anthology is the most widely quoted source for AIS costs. One cost that consistently appears in AIS reports is "$5 billion." First written two years after zebra mussels were found in Lake Erie, the federal Nonindigenous Aquatic Nuisance Prevention and Control Act of 1990 states: "Potential economic disruption to communities affected by the zebra mussel due to its colonization of water pipes, boat hulls and [infrastructure] has been estimated at $5,000,000,000 by 2000."

Although that estimate proved to be grossly overstated, a decade later the NCEE stated that "a number of reports and publications have reported that the cost of the mussels [is] around $5 billion." Five years later, U.S. Fish and Wildlife said that "for the 10 year period from 1990 to 2000 the cost of zebra mussels was about $5 billion." More than a dozen major reports after 2002 cite the 1990 estimated cost for zebra mussels as the actual cost between 1990 and 2000. In 2008 Tahoe newspapers used the 20-year-old estimate and reported: "U.S. Congressional researchers estimated that the Great Lakes zebra mussel infestation had an economic impact to industries, businesses, and communities of more than $5 billion."

Two years later, The *Quagga-Zebra Mussel Action Plan for Western U.S. Waters* says: "Congressional researchers estimated that mussel infestations in the Great Lakes area [had] an economic impact to industries, businesses, and communities of more than $5 billion." And the Tahoe RCD website still answers the Frequently Asked Question, "What is the economic impact of the Quagga and Zebra mussel?" with the same inaccurate cost.

The federal ANS Task Force now says the actual cost of the impact from zebra mussels in the Great Lakes from 1990 to 2000 was about $300 million, putting it in line with the Anderson Group's accountings, and about $1/18^{th}$ of what is generally claimed by AIS management groups and agencies.

The updated *Lake Tahoe Region AIS Management Plan 2014* says the potential annual impact from AIS could be as much as $78 million—more than twice what the entire nationwide economic impact from zebra and quagga mussels currently is. The 2015 *Implementation Plan for the Control of Aquatic Invasive Species within Lake Tahoe* states: "An analysis of potential AIS economic impacts to recreation, tourism, and property values, and increased boat/pier maintenance costs in the Lake Tahoe Region was estimated to be [up to] $78 million per year. However, these estimates do not capture the potential economic effects on ecological function for the sensitive and unique biological community in Lake Tahoe."

Currently seeking $4.5 million per year for 10 years from the federal government to underwrite research and to prevent and control aquatic nonnatives in Lake Tahoe, the destructive AIS invasion that in 2007 TRPA said was imminent throughout the Sierra Nevada, and has repeatedly restated since then, didn't happen in any of the mountain range's 2,000 lakes and ponds. And excluding the man-made Tahoe Keys canals and lagoons, the annual economic impact in the Tahoe-Truckee watersheds due to all aquatic invasive species remains the same as it was when John Fremont ascended the Truckee River and discovered Lake Tahoe —nothing.

Steve Urie

— 12 —

TRPA: Too Important to Fail?

[TRPA is] an agency of bureaucrats running amok.
  ~ Sharron Angle, Nevada Legislative Minority Whip

Sharron Angle's harsh criticism of TRPA was made because she believed the agency's building policies to be overly restrictive, and her assessment appealed to her constituency. Pushing an emotional button sure to incite less-government advocates, in 2003 Angle introduced a Nevada Assembly bill to withdraw from the TRPA, a measure that would dismantle the agency and end many meaningful government services at Tahoe.

Angle reignited an issue that had been unsuccessfully introduced in the Nevada Legislature five times before. Eight years later in 2011, a similar bill passed. Ed Gurowitz, an Incline Village management consultant, had partially sided with Angle's views and had written in 2005: "In the long run, I believe that the only recourse we have is for Nevada to secede from TRPA and regulate ourselves."

However, Gurowitz' agenda was to do what was best for the bi-state Tahoe community. A Californian who had lived on the Nevada side of Lake Tahoe for a decade, Gurowitz understood the Basin's ideological divides and political complexities, and he wrote a political column for the *North Lake Tahoe Bonanza*. In 2011 when Nevada was debating whether to withdraw from the TRPA, he reversed his position and wrote in the agency's defense, "We can't afford to let a group of people who think their political ideology and short-sighted opinions and interests are more important than the good of the rest of us and the health of a lake that belongs to us all."

Gurowitz' views had changed, and he embraced the necessity of an agency that he had earlier thought was out of touch. In July 2011 he accepted TRPA's offer to become their part-time Chief Operating Officer. Upon taking the job, he wrote in his parting *Bonanza,* column "After writing about the Tahoe Regional Planning Agency critically during the Juan Palma years, hopefully during the John Singlaub regime, and enthusiastically as I've watched Joanne Marchetta engage with reinventing the agency, I've decided to put my money where my mouth is."

Before assuming his TRPA position, Gurowitz wrote: "I am looking forward to working with people in the agency and to continue to turn around the image of the TRPA in the community—to have it be more customer focused and say yes rather than no." With the future of the agency on the line, he was brought in at exactly the right time.

In a 2005 *Bonanza* column, Gurowitz had written: "I moved to California in 1981, when the state was just coming out of one of its periodic water shortages. Shortly after I got there it began to rain, and it rained hard and long. Yet no matter how much it rained, I never saw any government official say that the drought was over. Their statements were more on the order of 'things are better, but we still have a long way to go.' Even when trees and homes were being washed off hillsides, their story never changed. Then I noticed that all those officials had the word 'drought' in their title—and I realized that if they were to declare the drought over, they would be declaring that their job was unnecessary. This was an epiphany for me about bureaucracy."

During his year as TRPA's part-time COO, Gurowitz saw his insight into bureaucracy in action. Threatened with extinction, TRPA aggressively responded to Nevada's threat to shut it down and focused on completing a revised Regional Plan that updated environmental standards, streamlined permitting, and promoted economic stimulus measures. In August 2013, California and Nevada's governors agreed to the plan update, and Nevada remained in the bi-state compact.

Sitting crossed-armed at the 2013 Lake Tahoe Summit, California Governor Jerry Brown reflected most Californians' feeling about the updated plan he had agreed to just days before. Others were more outspoken. Laurel Ames of the Coalition to Protect Lake Tahoe wrote in the *Sacramento Bee* that the "plan would strip Lake Tahoe of protections." David von Seggern, chairman of the Toiyabe Chapter of the Sierra Club said, "The plan doesn't do justice to the lake." And a California attorney for Earthjustice told the *Los Angeles Times*: "We got rolled." Many Californians were annoyed because between 1997 and 2013, of the $1.74 billion that flowed to TRPA's environmental programs, they had contributed 38% of the total, Nevada had contributed 6%—but by leveraging the threat to withdraw, the Silver State had gotten what they wanted—and the federal government, who has total environmental regulatory control in the Basin, only contributed 33%.

Gurowitz wrote: "The TRPA Regional Plan Update brought about unprecedented collaboration between the two states to regulate development in the Tahoe Basin while ensuring the security of the environment." When his contract ended, Gurowitz periodically contributed to the *Bonanza*. In his first column after returning, he wrote: "I still work with TRPA at about 30 percent of the time, doing staff training, development, and coaching. I'm committed to the agency's vision, mission and goals, and feel that the people who work there are the best of the best in the environmental field."

TRPA may hire superior staff and their training may be exemplary, but contrary to Gurowitz' claim, their ecologists and environmentalists aren't the best of the best, and there aren't very many of them. In January 2015, the Aquatic Resources Program Manager was the only staff member listed under the Environmental Improvement Program who was solely associated with the AIS programs. The agency has no research facilities and relies on UNR and TERC researchers to guide TRPA staff's science-based decisions and recommendations to the 15-member Governing Board that determines the agency's policies.

As critical as Tahoe's protection is, the federal agency has proven itself ill-equipped to deal with elementary aquatic policy issues. They and Tahoe RCD ignored the most definitive guides for determining AIS risk and the allocation of management resources: Whittier's, *A calcium-based risk assessment for zebra and quagga mussels,* and the U.S. Bureau of Reclamation's *Mid-Pacific Region Dreissenid Mussel Infestation Susceptibility Assessment,* which confirmed the Whittier study.

Certainly, no one from TRPA researched freshwater mollusks, other than to accept what they were told by UC Davis and UNR researchers, who received lavish grants from the agency to confirm TRPA's AIS policy and programs; nor did anyone read the studies of outside experts, such as Gerald Mackie and Renata Claudi. Instead, TRPA created the nation's most aggressive—and least needed—boat inspection program and implemented incredibly environmentally destructive and financially wasteful AIS control programs. And excluding game fish, they institutionalized a highly prejudiced attitude towards the few introductions that have been able to establish in Lake Tahoe.

Environmental advocacy is almost always beneficial, and it's appropriate to error on the side of caution when charged with protecting a national treasure such as Tahoe. But when advocacy turns to simple bullheadedness over refusing to admit failure or retract disproven science, the public and their agencies are misinformed and poor policy results. Arguably, the Asian clam control program that suffocates all benthic life under acres of pond liner is the most ill-conceived AIS control program in history.

Before the start of the Emerald Bay project, researchers told the funding agencies that the barriers would be removed in the fall of 2013. In early 2014 after the rubber matting was left in place for a second year, TRPA was asked why. Aquatic Resources manager, Dennis Zabaglo avoided the question and wrote that that "preliminary results were encouraging." Asked to see the results, he deferred to TRPA's Public Information Officer Jeff Cowen, who wrote: "If you are looking for interim or preliminary monitoring and data, you might check with the research institutions that are carrying out the work. TRPA typically only receives completed reports." That, of course, brought into question what preliminary results Zabaglo was referring to when he said they were encouraging.

Pushed for data from the Emerald Bay project, Cowen provided a link to the U.S. Forest Services' webpage for projects supported by money from the Southern Nevada Public Lands Management Act, which had granted $320,971 to the project. The deeply buried webpages are a trove of grant information and data on Tahoe water quality and environmental studies, and is the source for much of the information on AIS studies described in this book. Apparently neither Cowen nor Zabaglo, nor any of their colleagues were aware of the researchers' report, or they didn't attend meetings of the Asian Clam Working Group, because the report says that all data from five 2013 Asian clam population surveys was reported to the working group.

Cowen also sent graphs showing that until Tahoe water temperature reaches 60°F, dissolved oxygen under the rubber mats in Emerald Bay remained high enough to sustain the clams, and that in some areas there was only partial mortality when that temperature was exceeded. In those areas more than a third of the clams' eggs were able to develop into veligers—clearly, enough offspring to sustain their populations. One of the studies linked from the Forest Services' webpage was a report co-authored by TERC and UNR that TRPA's AIS managers had also apparently failed to read. It concluded that suffocating Asian clams under rubber mats was ineffective. (Wittmann 2013)

TRPA's AIS policies and programs were formed, validated, and enabled by three men: Sudeep Chandra, Geoffrey Schladow, and Ted Thayer. In 2008 Thayer became the face of TRPA's AIS program and remained so for the next five years. Promoted from Wildlife Program Manager to Natural Resource and Science Team Leader, he was TRPA's spokesman and liaison for implementation of the agency's AIS prevention and control programs. In 2009 he led AIS media coordination and organizational efforts to evaluate AIS threats and develop solutions. In 2010 he was promoted to AIS Program Manager. In 2013 he was named Implementation Coordinator for the Environmental Improvement Program, and in mid-2014, he was made division manager of TRPA's Research and Analysis Division. A year later, he resigned from TRPA.

Asked in 2012 what the "noticeable impacts" and "threats to water systems" from Asian clams in the lake were, Thayer avoided the question and wrote: "Once data is available on the current Emerald Bay project, managers will be at the point of producing the long-term strategy for addressing control of Asian clams. Strategies will be evaluated in the context of current and future funding." The program eventually died, not because of funding, but because it didn't work—a result that was predicted by many before it began.

Pressed on the statement, "It seems that the Asian clam was in Lake Tahoe for many years before there were noticeable impacts on the ecology and the recreational values of parts of the lake. If left unchecked these impacts are likely to have increased." Thayer wrote: "I will try to get you some additional information in the future. Unfortunately I have several other items on my plate right now, so I cannot promise when I will be able to dig up and send you anything else. As I mentioned, the managers in the [Asian clam] working group will be going over the various datasets to develop draft strategies once we have results from the Emerald Bay Project. We will have more information at that time."

After the rubber mats in Emerald Bay were rolled up and stored, he was asked to review a draft of this book for accuracy and comments. After no communication for more than two years in February 2015 he wrote: "I no longer work in the AIS program. Please direct any further questions that you have about the program or specific projects to our Public Information Officer." Four months later, he no longer worked for TRPA.

Geoffrey Schladow designed and oversaw the Asian clam control program that spread acres of pond liner across the bottom of Lake Tahoe. Although he declared the program a "scientific success," from every aspect, it was a total failure, and the long-term strategy that two years earlier Thayer said would be developed upon project completion was to store the mats and investigate alternative methods.

Schladow received a February 2015 draft of *Tessie*, and he was asked about his claims that the Emerald Bay Asian clam program was successful when by any criterion it failed, and why he continues to seek contributions and public grant money for research into new control methods when his studies show that Asian clam control in Lake Tahoe is impossible. He was also asked to comment on Cornell University's criticism of his Asian clam studies and his program recommendations to TRPA. He remained silent and offered no defense for his science or programs.

But there is hope that current attitudes are moderating. in July 2014, Schladow's UC Davis colleague, Katie Holzer, wrote: "Attempts at invasive species management that don't succeed in the long run may simply be a non-ideal use of limited resources, but there is also increasing evidence that many intensive management actions have the potential to directly or indirectly harm native species." Holzer may very well have been writing about the Asian clam control program. It is this balanced view that has been absent from Tahoe's AIS management programs.

For nearly a half-century aquatic science at Lake Tahoe flourished under Charles Goldman's direction, and a strong public trust was built with Tahoe's aquatic scientists. That trust is unravelling. As often happens when there is too much money to solve too few problems, waste and poor policy results. Tahoe doesn't have any harmful AIS issues and isn't threatened by any alien aquatic animals. Yet since 2011, TRPA has unsuccessfully lobbied Congress for $45 million over ten years to continue their aquatic research, fund AIS control, and to pay for the half of the boat inspection program not funded by boaters.

Sudeep Chandra is the most influential advisor on TRPA's AIS policies. When justifying recommendations that aren't science backed, he says, "It is better to err on the side of caution." That is wise advice that should be taken before needlessly killing tens of thousands of fish and spending millions to kill most life on acres of lake bottom. He ignores science that disproves his hypotheses, he creates ambiguity where none exists, he manipulates data, he exaggerates and misstates test results, and he uses his university position to give credibility to his unfounded recommendations for the agencies that support him. Yet, his bogus quagga mussel and New Zealand mud snail studies remain the only justifying science behind Tahoe's boat inspection program.

Chandra's debunked studies and his false claims that Asian clams increase calcium concentrations, that crayfish, clams, and warm-water fish decrease Tahoe's clarity and water quality and cause algal blooms are shockingly ignorant of basic aquatic science and ecology. When his studies were increasingly criticized, Chandra turned his attention to insignificant research that can't be easily analyzed or challenged, but he persists with his alarmist views. In a January 2015 feature story in the Sunday *Reno Gazette-Journal,* he reported that since the 1960's populations of the Tahoe flatworm, stonefly, and blind amphipod have crashed, [probably due to intentional introductions] and exclaimed: "They are disappearing. It's unprecedented. It's absolutely dramatic."

In February 2015, Chandra was sent a draft of this book and asked to refute the claims that his science was deeply flawed and to acknowledge that many studies disprove his claims that New Zealand mud snails can survive in High Sierra lakes, that Asian clams and crayfish cause algal blooms, that Asian clams in Lake Tahoe cause a loss of biodiversity, and that their shells leach calcium into the lake's water, creating hot spots where quagga can survive. Instead of responding to accusations that his claims are false and his recommendations are ill-advised, he deflected the issue and doubled-down by making his remarkable public announcement that "adult quagga can live in Tahoe's water and that successful reproduction can occur and juveniles survive to repeat the cycle as adults"—a biological impossibility.

Although there is necessary overlap, public agencies are careful not to cross jurisdictional boundaries and avoid stepping into each other's domain, and there is a hierarchy of law, respect, and deference: federal agencies are at the top of the authority pyramid, municipalities are at the bottom, and state agencies are sandwiched in between. California and Nevada's 1969 compact created a unique federal agency that has near-total leverage and final say over land usage and environmental policy and controls in the Tahoe Basin.

Regional interests divide Tahoe's residents and property owners on issues from commercial and residential development to transportation. Beginning with the 2007 quagga mussel scare, AIS management programs comprised TRPA's least contentious policies. But that changed in 2015 when TRPA backed the Tahoe Keys Property Owners Association in using chemical herbicides to control weeds. The Tahoe Water Suppliers Association, whose members supply drinking water from the lake to six local counties, was the most vocal critic. TWSA Executive Director Madonna Dunbar said that because Tahoe's water suppliers don't treat for chemicals and the average "drop of water stays in the lake for 700 years before it leaves," she was concerned that chemicals could migrate from the Keys into the drinking water.

An aroused public flooded local media with letters and op-eds opposing the use of aquatic herbicides, a Stop the Tahoe Keys from Using Herbicides in Lake Tahoe Facebook page garnered 1,700 Likes in less than two months, and more than 1,000 people signed an online petition demanding that herbicides be prohibited. Instead of confronting the problem head on, TRPA punted the issue to the Lahontan Regional Water Quality Control Board and the EPA. Taking the issue to the EPA was deceitfully disingenuous. TRPA was created to resolve precisely this type of special environmental issue. The EPA isn't constructed to deal with local issues, and the only assistance they gave was to acknowledge that it is permissible to use approved herbicides.

Regulators are charged with protecting the public interest, and to effectively discharge their duties, they must maintain an arms-length distance from those they regulate. This is especially true when formulating technically complex policy and regulations, and science-based programs should be critically evaluated by independent experts. When distance isn't maintained, program efficacy suffers, public trust is diminished, and resources are wasted. TRPA's Asian clam control program is an example of how a few researchers and agency managers can squander a lot of public money, while doing more environmental harm than good.

The 2009 program for Development of Asian Clam Control and Monitoring Plan Strategies for Lake Tahoe requested funds for Asian clam research from the Lahontan Regional Water Quality Control Board. The grant request was co-written by UNR's Sudeep Chandra, TERC's Geoffrey Schladow and John Reuter, Tahoe RCD's Nicole Cartwright, TRPA's Ted Thayer, and U.S. Fish and Wildlife Service's Steve Chilton. The program illustrates how a coalition of credentialed researchers and agency managers can readily tap public agencies for funding for scientifically unwarranted and environmentally unneeded projects.

The Introduction to the 2009 grant request begins: "Asian clams are a non-native freshwater bivalve that has established in Lake Tahoe and is causing apparent associated environmental impacts. It has been observed at low densities since 2002, but recently (April 2008) populations have been discovered in much higher but patchy densities in the southern portion of the lake. ... Researchers and [various agency staff] recently formed a working group to prioritize research, monitoring, and control projects of Asian clam populations in Lake Tahoe."

The funding request for $588,720 was granted and another $916,495 was approved by other agencies. The entire body of supporting science to the 2009 funding resolution stated: "The Asian clam poses a threat to water quality and beneficial uses through transformation, concentration and excretion of high levels of bio-available nitrogen and phosphorus into the water column and substrate. Asian clam colonies contributed to nearshore blooms of filamentous algae in 2008 and 2009. These algal blooms negatively affected the aesthetic enjoyment of Lake Tahoe. Additionally, decomposition of Asian clam shells increases localized calcium concentrations in the sediment substrate, creating a more suitable habitat for potential establishment of invasive zebra and quagga mussels." Although all of the claims were inaccurate, it's easy to see how six board members with little expertise in aquatic biology, would have to rely on the recommendations of two universities and three public agencies.

Although the TRPA is a small federal agency, over its half-century of existence it has slowly begun to emulated the rigid, bureaucratic style of its giant sister agencies. While some argue TRPA is overly-protective of the Basin's aesthetics and intrudes on property owners' rights, others say the agency doesn't do enough to protect the Basin's environment and has squandered vast sums of money. Sadly, both accusations are true. TRPA is poorly structured to manage the environmental complexities in today's Tahoe Basin. Nowhere is that more evident than with their AIS prevention and control programs. Instead of following guidelines provided by other federal agencies, TRPA deviates from mainstream science, performs small-scale studies that produce questionable results, and develops unique AIS programs—none of which have been successful.

Since 2004, 361 Tahoe environmental restoration projects and research studies have been underwritten by grant funding from the Southern Nevada Public Land Management Act—but no breakthrough science, insightful ecological findings, or even acclaimed journal articles were produced by the academics and researchers, who received disproportionately generous grants based primarily on the illusory goal of assuring that Lake Tahoe remains blue.

The primary reasons that no new aquatic ecology has originated at Tahoe is because the lake has sparse aquatic life to study and there are very few water quality problems to resolve. When Secchi-depth clarity is 70-plus feet and an arbitrary goal of 97.4 feet is set, and the lake's water quality ranks it at the top of the world's purest lakes, it becomes necessary to contrive issues to justify environmental research and water quality programs.

Most science academics are under constant scrutiny from their peers, who they compete with for stature and grant money. However, the emotional reverence for Tahoe's clarity and TRPA's unique status among public agencies has created an uncommon situation where a steady stream of public funding flows for arcane and unnecessary research and unneeded projects.

The updated 2014 *Lake Tahoe Region AIS Management Plan* reports that "efforts are currently underway to control invasive plants, Asian clams, crayfish, bullfrogs, and warm-water fish. Research is being conducted to determine the most effective means of controlling each of these species." Before conducting more research on how to control benign aquatic animals and to avoid programmatic disasters similar to the Asian clam control program, it would be prudent to retain experts to determine if any of Tahoe's non-native aquatic animals are truly "invasive" and need to be controlled.

Steve Urie

— 13 —

Management Paradox

It's amazing how extensive the [AIS] indoctrination has been: "Non-native species are bad—we've got to get rid of them." Boy, if you want nature to stop, you're going to be miserable.

~ Mark Davis, Biology Chairman, Macalester College

A September 2013 *Los Angeles Times* article reported: "State fisheries biologist Dave Lentz poured poison into a remote High Sierra stream and watched quietly as every rainbow and golden trout in the water turned belly up. After the rotenone spread along 11 miles of Silver King Creek, other biologists poured in a neutralizing agent, making the river again habitable—and a suitable home for the rarest trout in the world.

"Kneeling beside a small brass spigot that dripped the milky white toxin into a pool edged with alders, Lentz, a conservation coordinator for native trout with the California Department of Fish and Wildlife, smiled and said, 'Looks like everything is working as intended. The Paiute cutthroat trout belongs in this stream, not the nonnatives in here now.'

"The Paiute trout is native to the Alpine County stream in the Eastern Sierra's Carson-Iceberg Wilderness of the Humboldt-Toiyabe National Forest. But it had been squeezed out by rainbow and golden trout, which are not native to this portion of Silver King Creek. The plan to restore the Paiute trout had been held up in federal court for more than a decade by opponents who believe that poisoning a stream is about the worst thing that could happen in a designated wilderness. They also question the safety of rotenone and worry about its possible long-term effects on wildlife and regional water supplies. Biologists say rotenone is a natural poison that poses no threat to water supplies.

"After the poisoning, biologists in waders sloshed in knee-deep water, using nets to scoop up the trout and place them in buckets. The remains were to be tossed into the forest as an unexpected banquet for insects, birds and mammals.

"Alpine County Supervisor Don Jardine is among the critics of this recovery effort by the state Department of Fish and Wildlife, U.S. Forest Service, and U.S. Fish and Wildlife Service. Jardine first fished this wilderness with his father more than half a century ago. 'I have a basic philosophical objection to anyone polluting a natural waterway,' Jardine said.

"Jardine also pointed out that rotenone is banned for use in U.S. coastal waters and banned entirely in Europe. Also, some rotenone treatments have flopped. In 1992, an estimated 1,000 trout were accidentally killed when state wildlife biologists mistakenly used excessive amounts of potassium permanganate to neutralize rotenone. In the late 1990s, state wildlife authorities laced the Sierra reservoir of Lake Davis with rotenone as part of an effort to eradicate northern pike, an invasive saw-toothed fish that had been ravaging the lake's trophy-size trout. The pike returned in 18 months." (Sahagun, 2013)

The 10-year battle to eliminate non-native trout and stock Silver King Creek with Paiute trout highlights not only Fish and Wildlife's illogical persistence to reengineer nature, but also the futility of some their fishery management practices. The greatest issue for the endangered Paiute trout was the hybridization and predation by the rainbows and cutthroats that Fish and Wildlife had unwisely stocked Silver King Creek with. It will be years before it is known if Paiute cutthroats take back the creek, but most anglers are betting that those non-native fish, whose eggs weren't killed by the rotenone, will soon return.

The University of California, Division of Agriculture and Natural Resources lists 115 species of freshwater fish in the state. Of those, 64 species are native and 51 are non-native (worldwide, except in habitats not influenced by development or agriculture the ration of native to non-native species is typically 50/50 ±20). For many fish and wildlife managers, that's 51 species of invasive fish that need to be at best eradicated and at worst controlled, and that is where conservationist, recreationists, and sportsmen clash. Ecologists such as Mark Davis say it's folly to attempt to manage nature, and others such as Sudeep Chandra say that if we don't manage nonnatives, ecosystems will be stressed, there will be a loss of biodiversity, and nature will be negatively impacted.

There are many examples of successful habitat restoration projects where native species were reintroduced, but there are virtually no examples of successful AIS prevention programs and few examples of successful AIS control projects. Balancing environmental harm against the cost of controlling nonnatives is difficult. AIS agency managers often say prevention of invasives is less expensive than control. When that statement was challenged at a meeting of a dozen wildlife managers and ecologists, no one could cite a single example of where it was true.

City of Portland ecologist, Toby Query states his evolving views on AIS: "I have slowly shifted my thinking from one that 'combats evil invasives' to a more nuanced approach that targets thresholds and moves the system to a healthier state with the lowest overall impact. Interventions to restore habitat need to better evaluate the impact on the ecosystem as a whole."

Because they understand it is the purpose of science to explain and expand awareness in their field, biologists accept being proven wrong, and when inaccurate science is corrected, it means that their area of expertise has advanced for the common good. Accomplished scientists are studiously careful to separate fact from theory. Consequently even after retesting, when researchers obtain results that are contrary to prevailing science they invariably defer from making recommendations, qualify their conclusions, and say additional research is needed.

In his *Quagga Mussel Risk Assessment,* Chandra said more research was needed to determine if quagga could survive in Lake Tahoe. But then without any qualifications, he leapt to false conclusions, made sweeping generalizations, recommended boat inspections, and stated a possibility that defied all previous research: "The possibility exists for adult quagga to survive, grow, and reproduce in the Tahoe environment. The assumption that oligotrophic water bodies low in calcium are at very low to low risk of quagga mussel invasion is not necessarily supported."

Phase one of Chandra's assessment established Tahoe's nearshore baseline calcium at 9.16 ppm—61 percent of the accepted threshold for sustained quagga reproduction. His study hadn't shown that quagga could survive, much less reproduce, in low calcium water. But because it wasn't published in a peer-reviewed journal, recommended an ultra-cautious approach, and affected an isolated situation, it didn't receive any attention from the aquatic science community, nor is it referenced in the literature on mollusks or aquatic invasive species.

Worldwide, there has been no peer-reviewed study that supports the conclusions and recommendations of the *Quagga Mussel Risk Assessment*. When Chandra's science was challenged, instead of providing supporting evidence for his conclusions, he deflected one criticism of his study by writing, "I feel sad for you if you are unable to accept the uncertainties of nature." After a decade of receiving generous grant funding from public agencies to produce supporting AIS science for their programs and policies, Chandra has yet to publish in a peer-reviewed journal, and he still doesn't understand that the very purpose of those grants was to remove the uncertainties of Tahoe's aquatic science, not blur it.

Occasionally to maintain credence for a science-based program, agencies knowingly retain disproven theories. This is especially true if a false premise is the basis for a funded program. Because New Zealand mud snails, spiny water fleas, and hydrilla do no harm, even if they could survive in the Sierra (which they can't), boaters would rebel if inspections were based on preventing them from invading Lake Tahoe. Consequently, quagga mussels are retained as the poster-child for Tahoe's AIS programs.

It's difficult for TRPA and Tahoe RCD to reverse their stance after collecting every Tahoe boater's money and telling them since 2008 that if so much as a single quagga larva found its way into the lake, they could wreak havoc on the ecosystem. And it's more difficult to acknowledge that research they should have questioned was instead promoted by them as fact. So TRPA doesn't admit their errors, refutes the need for an independent risk assessment, sticks with ill-conceived prevention and control programs, and praises themselves for keeping Tahoe mussel-free.

In a December 2013 press release, Tahoe RCD issued the results of the prior season's boat inspection program. Of the 7,000-plus boats that were inspected, six out of ten paid an extra $35 to be decontaminated because they were not deemed to be clean, drained, and dry. And "boat inspectors intercepted more than 35 vessels harboring invasive species." TRPA wrote, "We're very happy with the watercraft inspectors' diligence and accuracy. Monitoring confirms that quagga mussels have not established in our lakes. These results are a credit to the inspection program"—and to having cold, pH neutral water with extremely low calcium.

In March 2015 after the *Reno Gazette-Journal* article, TRPA Chairman Casey Beyer praised the success of the boat inspection program and wrote: "In 2009 TRPA ... launched the Lake Tahoe AIS Program and mandatory boat inspections to help prevent the introduction of quagga and other aquatic invasive species at Tahoe. Now recognized as a model for the country, the inspection program has overseen the certification and safe launch of nearly 200,000 boats at Lake Tahoe.

"Federal funding that has paid for about half of the boat inspection program's $1.5 million annual cost is running out. TRPA is working to secure state funding from California and Nevada to ensure that this vital program continues, and that inspection fees remain affordable for the thousands of recreational boaters who visit Tahoe each year. Boat inspections have successfully kept quagga out of Lake Tahoe and remain a top priority for TRPA.

"A major priority for TRPA is completing a comprehensive AIS control strategy. The goal is to prioritize control projects and guide our future efforts to best use limited funding for maximum possible effect. The strategy will look at what invasive species are the most harmful at Lake Tahoe and where, when, and how they should be treated. Combatting aquatic invasive species at Lake Tahoe is one of the top priorities for TRPA." Beyer concluded by writing, "The fight will be long and difficult, requiring constant reevaluation to ensure that projects are as effective as possible."

TRPA still can't point to a single successful AIS eradication project anywhere, much less at Lake Tahoe, nor can they identify a single aquatic species that threatens the health of the lake. Yet, in 2015 when it was announced that boat inspections would startup May 1, a press release stated: "Invasive species, such as quagga mussels, New Zealand mudsnails, and hydrilla, are known to multiply quickly and colonize underwater surfaces, including docks and piers, water supply and filtration systems, buoys, moored boats, and even the beautiful rocky shoreline. They reportedly destroy fish habitat, ruin boat engines, and can negatively impact water quality and the local economy, recreation, and ecosystem." All of the statements are at least partially true—only not at Lake Tahoe.

In 2013 journalist David Bunker asked TRPA if because of low dissolved calcium and cold water, quagga mussel infestation was improbable, Public Information Officer Jeff Cowen wrote: "No parts of Lake Tahoe's nearshore have zero risk of infestation. Based on water chemistry and temperature, the suitability of habitat conditions in Tahoe range from low to moderate risk for the establishment and growth of quagga mussels." The U.S. Environmental Protection Agency totally disagrees with TRPA.

Cowen added: "Adult quagga mussels from Lake Mead have been successfully raised in the lab using Lake Tahoe water at typical ambient water temperatures." No quagga mussel has been raised in a lab using Tahoe-quality water, or has an adult quagga been sustained for more than 51 days. Ted Thayer, Geoffrey Schladow, and three TERC researchers were copied on Cowen's email, and no one corrected the misstatements to the journalist.

Beginning in 2013, Tahoe AIS information was modified to include the threat of "unidentified species." Non-native plant issues were magnified, and blaming nonnatives for outcompeting and choking out native species was expanded to also blame them for decreasing Tahoe's clarity and causing ecological changes in the lake's nearshore. In October 2013 when the Tahoe Nearshore

Evaluation and Monitoring Framework project was announced, a UC Davis press release quoted Sudeep Chandra: "The introduction of aquatic invasive species has already produced some profound changes in the nearshore. Further establishment of aquatic invasive species in the nearshore has the potential to unravel the tremendous progress made toward protecting Tahoe's clarity."

Six weeks before the *Reno Gazette-Journal* announced the stunning results from Chandra's quagga survivability experiments, the front page of an *RGJ* January 2015 Sunday edition was emblazed with the 3-inch headline "DEEP TROUBLE." The article hardly seemed to justify the emphasis; it repeated Chandra's earlier assertion that "profound changes are occurring in Tahoe's nearshore," and his theory that crayfish are the primary culprits. Chandra claimed crayfish were eating bottom plants and robbing Tahoe's tiny bottom dwellers of favorable habitat; and that the loss of clarity is in part blamed on the "concentration of nutrients" in crayfish excrement that fuels algal growth. He didn't mention that the introduction of trout and salmon for sport fishing and mysid shrimp to feed the trout had a greater impact on Tahoe's nearshore ecology than crayfish, a species Chandra labels as invasive. He said researchers were still studying ways to fix the issues facing Lake Tahoe, but that the problems could be reversed. Some ecologists are fearfully anxious to see how Chandra proposes to manage the crustaceans.

Villainizing crayfish at Tahoe is relatively new. In 2005 the *North Lake Tahoe Bonanza* reported: "Director of UC Davis' Tahoe Research Group, Davis Charles Goldman has researched crayfish for 46 years. In 1969, as a gift to Sweden, the Department of Zoology at UC Davis donated 70,000 crayfish, because Lake Tahoe's crayfish were found to be highly resistant to a common fungal disease plaguing Sweden's native crayfish, Goldman said.

'We thought we would get the Nobel Prize (for that donation). In Sweden, eating crayfish is like eating turkey at Thanksgiving.'"

In 2011, Goldman's former TERC colleagues said crayfish "increase algal blooms, decrease native invertebrates, and [are] detrimental to Tahoe's clarity." Although crayfish had been in Lake Tahoe for more than a century, UNR researchers said in six-years their populations exploded and were quadruple Goldman's estimates (two years later, Chandra said crayfish may have increased by another 50 percent to 300 million), and UNR Media Relations put out a press release that began: "The University of Nevada, Reno's Sudeep Chandra, a leading Lake Tahoe scientist who has studied invasive species and limnology at the lake for 20 years, said issuing permits for commercial harvesting of crayfish at Lake Tahoe will help improve clarity at the pristine lake, as well as take away a food source for other invasive species that threaten lake clarity and ecosystems.

"'The Nevada Division of Wildlife is taking an important step with the harvesting permits,' Chandra said. 'Harvesting the invasive crayfish at Tahoe could have a positive impact on lake clarity, especially in the critical nearshore zone of the lake.'" Because crayfish scour the lake bottom, feeding on biowaste, they improve clarity and water quality, and many ecologists disagree with Chandra when he claims that crayfish are bad for the lake.

After the first year of legal harvests, the *Tahoe Daily Tribune* reported: "Regional outfitters who say fish like trout and mackinaw rely on crawfish for food are voicing concerns. Gene St. Denis started fishing Lake Tahoe in 1981. A trophy trout and light tackle specialist, he said 50 to 70 percent of the fish he reels in have crawfish in their stomachs. 'They're a significant food source for Tahoe's fish,' St. Denis said. 'Saying crawdads aren't a primary food source for the fish isn't good science.'

"But stomach content can sometimes be misleading, according to Chandra. 'The food located in a stomach offers only a snapshot of a diet and not necessarily the animal's long-term energy source. The scientific data Chandra has seen indicates that mysid shrimp compose the bulk of their caloric intake.'"

Mysid shrimp and crayfish were intentionally introduced into Tahoe's food chain to supplement sparse natural food for game fish, and the two crustaceans provide the bulk of food for large Tahoe trout. Saying that one part of the food web is more desirable than another is misleading—depending on season and climate, food sources for fish are in constant flux.

Besides zebra and quagga mussels, TRPA increasingly targets New Zealand mud snails as a species to keep out of Tahoe. Chandra claims the snails can survive in the lake, will compete for algae, decrease biodiversity, and negatively impact fish. In a 2011 survivability experiment: *Short-term survival and potential grazing effects of the New Zealand mudsnail in an uninvaded Western Great Basin watershed,* Chandra and four researchers analyzed the ability of New Zealand mud snails to survive in water from Lake Tahoe and a Reno, Nevada pond. Although reputable studies say that mud snails can live out of water for only two days, the introductory paragraph of the two-week experiment says New Zealand mud snails can live out of water for three weeks.

The UNR report for mud snail survivability states: "Ideally, the experiment would have occurred over more than 14 days, but the need for immediate information to assist with ongoing management decisions as well as funding limitations necessitated the short duration of the experiments." The report concludes: "In fact, results from this, as well as other studies in the region, led to the establishment of new guidelines for boat inspections at Tahoe to proactively reduce the likelihood of nuisance aquatic species introductions." It's perplexing that five Tahoe researchers put their names on easily discredited science, but it's as concerning that TRPA didn't catch the errors, promotes the study's results, and says that only quagga and zebra mussels are potentially more invasive to Lake Tahoe than New Zealand mud snails.

Like the mussels, mud snails can't establish in Tahoe, and similarly, a flawed, short-term study headed by Chandra put the snails near the top of Tahoe RCD's list of AIS that boat inspections will keep out of the lake. A May 2013 front page story in the *Reno Gazette-Journal* was headlined: "They're here!" Reno area media featured the news that regional environmental agencies had issued grave warnings after three New Zealand mud snails were found in the Truckee River. The most startling fact reported was that the snails had "made their way 7,000 miles across the globe and into a stretch of the Truckee River running through Reno"—highlighting the folly of trying to manage nature and controlling the spread of non-native species.

Except for being much smaller than most of the thousands of other snail species—adults are about $1/32^{nd}$ of an inch, and several dozen fit on a dime—and having attractive whorled shells, New Zealand mud snails are unremarkable. They aren't toxic and feed on phytoplankton and detritus. But there must be something to not like about the nonnatives. Nevada Department of Wildlife biologist, Chris Crookshanks, told the *RGJ* not about mud snails but instead said, "Invasive species are known to choke out native species." Many believed there had to be more front-page worthy news than three snails from New Zealand possibly threatening native snails in a river stretch of Nevada's high desert.

The *Tahoe Daily Tribune* wrote of the trio of mud snails: "'It's certainly bad news,' said TRPA Senior Wildlife and Fisheries Biologist Patrick Stone. 'Any time the proximity of an infestation moves closer, the risk of that animal making its way into the region increases.' The snails compete directly with other wildlife and threaten fisheries by depleting nutrients fish depend on. ... And there's no natural predator in the region to keep the population in check." There doesn't have to be a predator; mud snails can't establish in Sierra lakes or the California stretch of the Truckee River, and in the Nevada portion of the river, New Zealand mud snails cause no harm.

Mud snails in Reno present a conundrum for TRPA and Tahoe RCD. Chandra's study says "boat inspections are believed to be a deterrent to the spread of New Zealand mud snails," and "the most effective way to minimize impacts is to prevent or slow snail introductions to new ecosystems. Invasion vectors include boats or other personal equipment (fishing gear, boots, etc.)." But there are no trailered boats on the Truckee River, which means inspections wouldn't have prevented their introduction.

Because boat inspections won't stop the intrepid snails, Tahoe's paddle boarders, kayakers, rafters, canoeists, and anglers should hope that local agencies take the time to review the science before they require personal watercraft, fishing gear, kids' water toys, and wet swimsuits to be inspected. As irrational as that sounds, worse AIS regulations have been enacted. In October 2012, the National Park Service banned scuba diving in Crater Lake out of fear divers' gear would introduce quagga mussels—Crater Lake's calcium is less than 12 ppm.

A commonly used deception for justifying boat inspections is to cite irrefutable unrelated facts to give credibility to unproven theories or validate generalizations. Lecturing in a continuing education program at UC Davis, in May 2015 Chandra showed a visual of where Tahoe's boaters come from—a web of lines covering the entire Western United States showed lakes of origin. He ended his talk by saying, "There is overwhelming evidence that invasive species are largely transported by boats. These boats are moving non-native taxa from [these] lakes to Lake Tahoe.

"Since 2007 scientists have tried to make the case for a change in environmental policies that just aren't about the clarity and the watershed but the biological and ecological attributes of the lake. [And] there has been a concerted effort to use science-based information to develop policy to protect the lake," Chandra concluded. Nodding in agreement, no one knew to ask if any of the species being transported could survive in Lake Tahoe.

Tahoe's ultra-oligotrophic water is inhospitable for New Zealand mud snails. They much prefer warm, high-nutrient water over cold, clear High Sierra water, and mud snails can't establish in low-conductivity water. (Herbst, 2008) In freshwater systems, conductivity is related to calcium carbonate or water hardness. Water with low conductivity is "soft," or low in calcium.

Compiled by Fisheries and Oceans Canada, Table B verifies Herbst's New Zealand mud snail's physiological tolerances:

Table B: Physiological tolerances of New Zealand mud snails

Parameter	None	Little	Moderate	High
Conductivity (µs/cm)	<25	25-200	200-1,200	>1,200
Temperature (°F)	<32	32-59	60-68	>68

(Therriault, 2010)

Dissolved minerals increase conductivity, and the measurement is a useful gauge of freshwater quality. Clarity is a useful indicator of water pureness, but measurements of conductivity and dissolved oxygen provide a better indication of water quality. Studies show that lakes and streams with good mixed fisheries have conductivity that ranges from 150 to 500 µS/cm, a spectrum where there are sufficient nutrients and the water is unpolluted. Tahoe water conductivity is less than 100 µS/cm. The lake's extraordinary water quality explains why there were few native fish in Lake Tahoe and why sport fishing is still marginal except in the nutrient-packed flow from the Tahoe Keys.

Whereas the TRPA has no scientists in its AIS program, the California Department of Water Resources (CDWR) has two environmental scientists whose work focuses on quagga and zebra mussels. The CDWR manages the State Water Project (SWP), which is the world's largest publicly operated power development and water conveyance system. The water transport system that allows California to flourish funnels snowmelt from High Sierra streams with calcium below 5 ppm to Santa Barbara County, where it is above 90 ppm. Along the way it provides water to 23 million people and irrigates 750,000 acres of farmland.

The State Water Project includes 34 lakes, reservoirs, and water storage facilities, 20 pumping stations, five hydroelectric power plants, and approximately 700 miles of canals, including the 444-mile California Aqueduct. If determining public risk and potential infrastructure damage due to AIS seems like a daunting task for the TRPA, consider CDWR's responsibilities. To assure it properly discharges its duties, CDWR not only has its own dreissenid mussel scientists, but also draws on the resources of the California Quagga and Zebra Mussel Interagency Team, which is comprised of representatives from five state agencies.

In performing its mussel risk analysis, CDWR didn't rely only on its own resources to formulate its AIS policies and to manage AIS threats to the SWP; it commissioned a study by RNT Consulting, North America's foremost authority on controlling dreissenidae. For 25 years, the Great Lakes-based consulting firm has performed AIS risk assessments, conducted environmental analyses, and advised governments, companies, and utilities on AIS remediation. RNT's chief scientist Renata Claudi headed the study and risk assessment for CDWR. The consulting firm shares their research, but along with all other science that invalidates Tahoe's AIS management programs, TRPA, Tahoe RCD, and their researchers pointedly rejected RNT's findings for the CDWR.

Table C is from RNT Consulting's report for the California Department of Water Resources' State Water Project:

Table C: Criteria to determine levels of dreissenidae infestation

Parameter	None	Little	Mod.	High
Calcium (ppm)	<10	<16	16-24	≥24
pH	<7.0;	7.1-7.5;	7.5-8.0;	8.2-8.8
Mean Summer Temp. (°F)	<64	64-68	68-72	72-75

(Claudi, R., et al., 2011)

Because of the critical public interests managed by CDWR, the necessity to protect public assets and assure quality drinking water, yet balance those concerns against recreational usage and environmental issues, conservative margins of safety were built into the management thresholds used by CDWR for establishing their AIS program and watercraft inspection policies.

RNT says that concentrated calcium, in relationship to pH and temperature, is the critical factor in determining potential mussel infestations. If pH is low in high calcium water, mussel shells become thin and eroded. Conversely, if calcium is greater than or equal to 11.5 ppm and pH is above 7.5, dreissenidae may survive, but won't propagate. In Lake Tahoe, as in most open water lakes, pH is near neutral (7).

A complicating factor in quagga mussel lab experiments is that lake water in closed containers quickly produces algae, which causes pH to rise. Consequently, water chemistry must be closely monitored in the lab to assure that it is representative of the conditions being tested. Low water temperatures are also a major deterrent to dreissenid mussel colonization. In the High Sierra, the combination of neutral pH, cold water, and extremely low calcium combine to prohibit quagga mussel propagation.

Fourteen of the large lakes that supply water to the SWP allow recreational boating. Half have calcium levels that the agency deems "unsuitable habit for quagga and zebra mussels," and they don't have inspections. The seven SWP lakes that have calcium of 16 ppm or higher and allow boating have inspections. Differing with Sudeep Chandra who says, "It's not all about the mussels," the CDWR believes that quagga and zebra mussels are the only aquatic animals that threaten the State Water Project. (Weisser, 2012)

An example of carefully controlling test parameters for a quagga experiment was demonstrated by the CDWR in 2011. The agency performed a quagga survivability test in which researchers slowly ran lake water with calcium averaging 11.5 ppm from large tanks into smaller tanks that contained quagga. Quagga survivability was achieved for 90 days, but because algae build-up in the large reservoir tanks caused pH to elevate above 7.5, researchers self-invalidated their results and didn't publish them. It is this level of scientific integrity that has been missing in studies performed for TRPA.

— 14 —

New Directions

The great enemy of the truth is very often not the lie—deliberate, contrived, and dishonest—but the myth, persistent, persuasive, and unrealistic. ~ John F. Kennedy

After failing to nurse quagga mussels through 90 days in artificially warm, calcium-laced water from the Tahoe Keys, in February 2014 Sudeep Chandra said that he would present results from his ongoing quagga mussel survivability experiments in a public venue. Some hoped that Chandra would use Tahoe RCD's May 2014 AIS Forum to bring closure to his position that quagga mussels could survive in Lake Tahoe, or to at least present an update. The annual symposium is the Basin's major AIS event, and it brings together the region's agency managers and interested public. The Tahoe Center for Environmental Sciences hosted the 2014 forum—the perfect venue to present new findings.

Because the conference was in the same building where Geoffrey Schladow has his office, it was also anticipated that he would provide an update on the high-profile Asian clam control project in Emerald Bay and what had been learned from the project. Chandra didn't attend the forum, nor did Schladow or any other TERC or UNR researcher make a presentation, nor was any new science presented. Instead, the audience heard how divers had once again cleared Emerald Bay of Eurasian watermilfoil, making Tahoe's least accessible beaches even more aesthetically appealing. A manager from Trout Unlimited claimed that anglers were those most responsible for the spread of New Zealand mud snails. And a volunteer weed control manager from the Tahoe Keys Property Owners Association explained that because the Tahoe Keys' total property value was more than $1 billion, the Keys were integral to Tahoe's economy. While displaying an aerial image of the tightly-packed development, he took a glass-half-full approach and pointed to the narrow strips of remaining marsh that he said "still filters sediment from the Upper Truckee River"—the projected photo plainly showed the broad plume of algae and sediment-laden goop that flows from the Keys.

Providing no new data, TRPA's Dennis Zabaglo briefly referenced the Asian clam project in Emerald Bay, but quagga mussels were ignored until the question and answer session, when a written question was asked of the "most qualified person to comment ... whether boat inspections are unnecessary because the calcium levels in the lake are too low to support a mussel population." Zabaglo stepped up and answered: "While calcium levels in the middle of the lake are extremely low, in discreet and isolated areas, such as marinas and the [Tahoe] Keys for example and around Asian clam beds, calcium is at a level where they can establish. ... We know they can live in Lake Tahoe water, but whether or not they can reproduce ... is among the research going on, but what we've been told by the researchers [is] that in these smaller areas and pockets they can more than likely establish."

Tahoe RCD's Nicole Cartwright followed Zabaglo and said, "The mussels aren't the only thing we are concerned about. And so while mussels may never get all over the entire lake and be able to survive in every nook and cranny, we are worried about everything. And the boat inspection program is going to prevent all introductions."

A year later in May 2015 when Tahoe's seventh year of mandatory boat inspections kicked off, in an interview with

Reno's Channel 2, Cartwright repeated some of the most common myths regarding Tahoe's aquatic nonnatives: "Prevention is much cheaper than any kind of control or eradication if new aquatic species would make their way in. The goal is to keep [quagga] out of Lake Tahoe. They would have a lot of effects on our recreation, on our economy, and just on our local ecosystem ... We wish we would have had the [inspection] program prior to the Asian clams getting here."

Regardless of how TRPA and Tahoe RCD frame it, there is no nook or cranny in Lake Tahoe that can sustain quagga, and there is no evidence that boat inspections prevent non-native introductions; at best, they slow their spread. Cartwright may be correct when she says there will be no new invasive species introductions in the lake, but it's not because of boat inspections—it's because the habitat is inhospitable to most aquatic species.

After the 2014 July 4th weekend a TRPA press release patted Tahoe RCD and themselves on the back for the fine job they were doing in protecting Lake Tahoe from AIS: After Tahoe RCD inspectors found "a boat with quagga mussels and an unidentified snail species hiding in the anchor locker. The boat was fully decontaminated ... and cleared to launch. 'The fact that this boat was predominantly clean, drained and dry, yet inspectors still found the mussel encrusted in mud on the anchor, is significant and proves that the rigorous Lake Tahoe Watercraft Inspection Program is working,' said Dennis Zabaglo."

TRPA Executive Director, Joanne Marchetta used the incident to appeal for funding for the boat inspection program. A week after finding the mussel and snail, she wrote: "Federal funding is drying up and the program is in urgent need of new revenue sources to continue boat inspections in 2015. We are tasked with a seemingly insurmountable challenge: how to continue to share this special place while also protecting it from the threat of infestation from AIS." Once again, TRPA used the thinly veiled threat of closing Tahoe to boating if money wasn't found for inspections.

The *Lake Tahoe Region AIS Management Plan* explains: "Oversight for state AIS management plans is typically led by a respective state resource agency; however, in the case of bi-state or regional plans, oversight is best suited to an organization capable of regulation across state jurisdictions. The TRPA has such regulatory authority. TRPA has successfully demonstrated the ability to lead and manage the $1.1 billion Environmental Improvement Program (EIP). Therefore, the TRPA will act as the fiscal agent for funds associated with implementing this Plan." In the last decade, no state has spent as much money on AIS management as TRPA has at Lake Tahoe. The agency's *2008–2018 Environmental Improvement Plan* calls for treating 4,000 invasive species sites along the 76 miles of the lake's nearshore—fortunately, the plan is falling far short of its goals.

Marchetta went on to write: "We've kept new destructive invaders out of Lake Tahoe over the last five years. ... However, our work is far from done. The stakes are high. To fail could mean unimaginable environmental and economic consequences. ... So, our focus must remain on both preventing the introduction of quagga and zebra mussels and other invasive species, while working to successfully manage those aquatic invaders, such as Asian clam. We will continue improving the inspection program and remain open to constructive feedback and suggestions."

The month before Marchetta's op-ed, she had been sent a draft of *Tessie, Quagga Mussels, and Other Lake Tahoe Myths*. She didn't acknowledge receipt of the book, nor did she reply to a letter asking for a meeting to discuss Tahoe's AIS prevention and control programs, nor did she respond to the suggestion to have independent experts perform an AIS risk assessment—and her staff ramped up the threat of aquatic nonnatives to Lake Tahoe.

In an April 2016 interview with the *Sierra Sun/Bonanza*, Marchetta said that AIS were one of TRPA's top five priorities, and that by 2020, she "hoped that in our AIS program that we've solved prevention and continue to have no new invasions of harmful species; and that on the control question—control of existing species that are already in the lake—that we have found funding for and found new innovative ways of reducing and, to whatever extent we can, eliminating existing invasions." It is a near certainty that no new AIS will invade Tahoe, and if TRPA finds a way to effectively control watermilfoil, curlyleaf pondweed, and warm water fish—species they've determined are controllable—by not using non-toxic chemicals, they will achieve a first.

In a July 2014 interview with *Reno News & Review*, Julie Regan, TRPA's External Affairs Chief was quoted as saying the "long-term consequence [of quagga mussels] is a huge negative impact on the fishery, on the food web of the lake as a whole, and of scenic and resource degradation where they attach to piers; they attach to any structure. ... We have a wedding industry in Lake Tahoe, so imagine people trying to have their wedding on a beach that is littered with degrading quagga mussel shells."

In scientific parlance, results recorded under uniform conditions and confirmed by multiple tests that are consistent with physical observation is a "law." A set of scientific facts, based on reliable principles that point to a probable conclusion, is a "theory." And a logical explanation of an unproven idea is a "hypothesis." To serve political agendas, Tahoe's science advisers have ignored scientific laws and mixed hypotheses with theories.

A basic scientific requirement is that when evidence is repeatedly at variance with a hypothesis or theory, it must be abandoned. Some hypotheses and theories that researchers have passed off as fact and deserve to be jettisoned include: Asian clam beds increase dissolved calcium; clams and crayfish cause algal blooms; New Zealand mud snails can establish in Lake Tahoe's low-conductivity water; spiny water flea are harmful; unchecked, non-native plants will spread across Tahoe's rocky bottom; warm-water fish harm biodiversity; and boat inspections prevent the spread of non-native species. And there is no Tahoe myth that is more contrived, persistent, and unrealistic than the one that repeats that quagga mussels can establish in Lake Tahoe.

No one wants to jeopardize Tahoe's ecology or aesthetics, but after years of outreach programs and spending tens of millions on AIS prevention and control programs, few Tahoe Basin residents and visitors can tell the difference between an Asian clam and a quagga mussel, or know which one has been quietly living in Tahoe for years and which one couldn't survive there for more than a couple of months. Very few understand the biology and habitat requirements of Asian clams, quagga mussels, or New Zealand mud snails, but almost everyone believes they should be kept out of Lake Tahoe and controlled if one sneaks in.

Geoffrey Schladow told the 2008 California Colloquium on Water that "it's going to require a lot of money to reengineer the Basin, and [to obtain the] money is going to require political will to do what the electorate, the stakeholders, want." What do the electorate and stakeholders want? Schladow says there is a "growing willingness to do whatever is necessary to protect Lake Tahoe." He's probably right. The biased and often inaccurate data and information that has been given to the TRPA, and what the agency's staff has selectively magnified and relentlessly passed on to the public says Lake Tahoe is potentially in great peril from aquatic invasives—a totally erroneous assertion.

Unwittingly, TRPA may be unravelling the half-century of responsible environmental policy and management they knitted together by taking up complex ecological issues they only superficially understand. It's their mission and duty to plan and administer Tahoe's land use and to protect its environment. As importantly, TRPA is the funding agency and provides oversight for the $100 million that has been annually distributed for more than 17 years by their Environmental Improvement Program.

TRPA doesn't have the personnel to internally formulate effective AIS policy, and they have ignored proven science and failed to properly evaluate questionable recommendations from a small circle of professors and researchers. And their misguided AIS prevention and control programs are undermining their successful habitat restoration and forest management programs. Because of Tahoe's high-profile image and their total control of the Basin's environmental policies, TRPA has legitimized discredited science, fostered unfounded AIS fears, and given credence to unwarranted management programs from California to New York.

Everyone wants to protect and preserve Lake Tahoe, but substantial questions remain regarding the need to "reengineer the Basin" and "to do whatever is necessary to protect it." If that were true, the Upper Truckee Marsh would be restored to its historical wetlands habitat; Asian clams wouldn't be suffocated under rubber mats; and the boat inspection program would be re-examined. It is time to include independent experts, draw on resources from agencies such as the U.S. EPA, and reformulate Tahoe's non-native aquatic species policies.

Most understand that protecting native species begins with preserving their habitat. Often it wasn't realized until years later that environmental mistakes such as destroying wetlands, clear-cutting forests, salting roads, indiscriminately spraying insecticides and herbicides, over-fertilizing, damming streams, or destroying natural drainage ponds were environmentally harmful. Today, we are careful not to repeat past errors, and habitat protection and reconstruction are receiving overdue resources.

For more than a half-century, the environmental message has been "Keep Tahoe Blue." The League to Save Lake Tahoe's slogan is interpreted in many ways. But Mark Twain may have said it best when he described Lake Tahoe's water as "not merely transparent, but dazzlingly, brilliantly so." Arguably Tahoe's water isn't as "pure and fine, bracing and delicious" as when Twain wrote those words nearly 150 years ago, but it is still spectacular, and most realize that the lake's eutrophication and water quality decline can be mitigated.

TERC's director said, "I think most people associate Lake Tahoe with its clarity and efforts to restore it." If Geoffrey Schladow's observation is correct, a growing segment of Tahoe environmentalists fear for the Basin's future. Tahoe's clarity is primarily a function of suspended microscopic particulates. The destruction of the Upper Truckee Marsh is the primary reason that tons of fine sediment are no longer filtered before mixing with unnatural levels of algae that are nurtured in warm Tahoe Keys water before flowing into Lake Tahoe.

Intuitively, most people understand that clarity is an indicator of water quality. But water resource managers know that there is much, much more to protecting the quality of Tahoe's water than improving the depth at which a submerged dinner plate can be seen. And that if the water Mark Twain drank is to be as pure and fine, bracing and delicious in another 150 years, meaningful environmental programs have to be instituted —but first, plans to reengineer the Basin and wasteful and destructive programs have to be scrapped.

After Notes

Some people think that the truth can be hidden with a little cover-up and decoration. But as time goes by, what is true is revealed, and what is fake fades away.

~ Ismail Haniyeh

In the fall of 2012, I wrote to Geoffrey Schladow and copied Ted Thayer: "I'm particularly interested in the growth rate and the specifics when you say: 'It seems that Asian clams were in Lake Tahoe for many years before there were noticeable impacts on the ecology and the recreational values of parts of the lake. If left unchecked these impacts are likely to have increased and possibly started to impact things such as water supply from the lake.' Please forward relevant data and reports."

Even though an Asian clam population survey had been recently completed, Schladow didn't reply and Thayer wrote that he had more pressing issues to manage and that he couldn't commit to when he would get back to me—he never did. When the director of an institute that is supported by contributions and public money and the manager who oversees the programs won't send data to back-up their premises, it's a good indication that their claims are questionable. And it was their lack of response, along with Sudeep Chandra's outrageous claims about the threats and harms that Asian clams cause, that prompted me to dig into AIS science and Tahoe's prevention and control programs. After a year of intensive research, it was apparent that using rubber mats to kill Asian clams wouldn't be effective, and that in fact the mats were ecologically destructive.

When the Emerald Bay project dragged on for a second year and Thayer was promoted out of the AIS program, I began an exchange with TRPA's Public Information Officer, Jeff Cowen. Over three months and more than 20 emails, Cowen sent some of the data I was looking for. It showed that the Emerald Bay project wasn't effectively eradicating clams and that the optimistic reports about the project given at the 2013 Lake Tahoe AIS Forum and that TRPA's Dennis Zabaglo gave to me in 2014 were simply wrong. Cowen also supplied links to U.S. Forest Service grant reports that were useful in researching UNR and TERC studies.

In early 2013, I became involved in the Truckee AIS management program. Teresa Crimmens had been hired as the Truckee Tahoe RCD manager, and she had guided the passing of a town ordinance that identified quagga and zebra mussels as the invasive species that would be prevented by boat inspections. When data was presented to town staff showing that quagga couldn't survive in Donner Lake, inspections were deferred for a year to enable a working group headed by Crimmens to examine all species that could potentially pose a threat to the lake.

I believed I was a contributing part of that process, and I summarized the research that I had compiled and sent it to Crimmens. She didn't acknowledge receiving my report, and soon after I sent her my data, the Truckee *AIS Vulnerability Assessment* was released. I wrote to her of my concern that the assessment was incomplete and needed work. She replied, "The Vulnerability Assessment is a draft. The intent is to collect comments/feedback at the Working Group meeting and make edits as necessary before finalizing." Two weeks later, she presented the assessment to the advisory group. At the conclusion of the meeting, ignoring critical comments, she dissolved the group. No significant changes were made to the draft assessment, and four months later the town council implemented mandatory inspections.

When it became apparent that Crimmens and Tahoe RCD weren't interested in evaluating whether the species they claimed were invasive to Sierra Nevada water bodies and were in fact harmful, or if they could even establish in local lakes, I decided to write this book. I had naively believed that presented with contradictory data from authoritative sources, responsible agency managers would reconcile the information and data, and if necessary, recommend appropriate policy changes to modify programs. It soon became obvious that wouldn't happen.

Following up on the $338,000 grant that Sudeep Chandra received in 2010 to study quagga survivability, in January 2014 I wrote to him: "Last May [2014] you wrote to the Town of Truckee: 'I am sure you understand the pilot study we conducted with adult quagga mussels using Tahoe water presented a set of conclusions that [were a] surprise to me. This is why we have requested funds to conduct an experiment to examine the survival across two life stages of quagga mussels (veliger, adult) with a reproductive analysis of adults if they survive the time allotted for the experiment. I am looking forward to sharing with you our final results from these new quagga survival experiments we are conducting with Tahoe water later in the year.'

"Have the experiments been concluded, and what were the results?"

Chandra wrote back that the experiments "have not been concluded due to delays in the project but we are starting our second set of experiments in March [2014]." At the time, I was confused why delays would require a second set of experiments, but months later when I read in his report to the Forest Service that 19 of 20 mussels died within 20 days in the first tests, his second attempt made sense. Chandra didn't accept my invitation to meet and discuss his studies and invasion ecology, nor did he send data to counter science or facts cited in this book. But he did write that I had made "many incorrect statements related to understanding ecology and nature [and] the work from [his] lab."

At the time I was piqued that without stating why, a research professor would avoid a question that he had promised to answer and only say that I was wrong. After investigating his research, I understood that it wasn't I who was wrong.

In 2007 when Lake Mead was found to be teeming with quagga mussels, instead of examining the science, TRPA passed a resolution declaring an emergency in their attack on aquatic invasives. Within months, boat inspections were instituted and Chandra began an eight-year campaign of misinformation and false science to justify the war against Tahoe's non-native species.

Compared to worldwide environmental problems, such as climate change, deforestation, natural resource depletion, ocean acidification, pollution, and over-population, killing most life on several acres of Lake Tahoe lake bottom and inconveniencing boaters are minor issues. However, that doesn't excuse those whose job it is to formulate and administer Tahoe's aquatic environmental policies. Dozens of technically trained organization and agency managers are involved in implementing Tahoe's AIS programs. I talked with many of them, and most were well-meaning, but almost all, without questioning the science, dutifully backed TRPA's misinformed AIS policies and repeated the false science fed to them by Sudeep Chandra's team of researchers.

The greatest bias in science is not people knowingly pushing an agenda—it's towards importance. Everybody who spends their hours, days, and weeks—perhaps even careers—researching something wants it to be important. It becomes incredibly difficult to not elevate the subject in your own mind, and to skew conclusions. Fortunately in science, those who reach the top of their profession do so because their research is insightful and often contributes to the common good.

More than 40 agencies and organizations are represented by the Lake Tahoe AIS Coordination Committee. Are all of the agency representatives stupid and uninformed? Of course not; they are just deluded, as we all are, by wanting their position to be more important than it is.

The unifying factor that each of its representatives shares is that their position is enhanced by having aquatic nonnatives threaten Lake Tahoe, and the greater the threat, the more important their work becomes. Ed Gurowitz observed that people with the word "drought" in their job description have their jobs jeopardized when it frequently rains. And when there is no environmental harm posed by non-native species, the same is true of those with "aquatic invasive species" in their job title.

Science academics and their labs are largely supported by public grants, and there is pressure on them to generate revenue for their departments. Unfortunately, the economics of university studies sometimes gets in the way of good science. In the three years between 2008 and 2011 when AIS research at Lake Tahoe exploded in efforts to rationalize the threat of non-native aquatic species and to justify the TRPA boat inspection program, Marion Wittmann, a post-doctoral researcher, who was then working for TERC, was either the principal or co-principal investigator in eight studies on Tahoe's aquatic species. The studies were funded by five public agencies and totaled $1,385,633 in grant revenue.

Wittmann's prodigious output is rife with contradictions, bias, and faulty science. In the summer of 2014, she was working as an associate at the University of Notre Dame. While gathering information on Tahoe's Asian clam control programs, I requested to meet with her in South Bend. She didn't reply to my request. In spring 2015, she became a researcher under Chandra's direction in UNR's biology department.

TERC researcher Brant Adams succinctly explained the TRPA-academic researcher relationship when he was asked by a TERC docent trainee, "What interaction does TERC have with

TRPA?" Adams replied, "A lot. They have the money, and we're trying to spend it."

For eight years starting in 2007, Sudeep Chandra, Geoffrey Schladow, and Ted Thayer were responsible for the formulation and implementation of Tahoe's AIS policies and programs. When I first attempted to establish a dialog with them, they were receptive to my questions. However, as soon as I pointed out errors in their science, they stopped replying to my inquiries and directed me to TRPA's public information officer.

Initial justification for Tahoe's boat inspections was based entirely on the *Quagga Mussel Risk Assessment*. When its flaws are exposed and its recommendations questioned, the response is that is why follow-up studies are being performed. However, other than incomplete interim reports from Chandra's fallacious 5-year study, in the six years after his quagga risk assessment, no additional research results to support his 2009 risk assessment were produced, and every attempt to demonstrate that quagga could establish in Lake Tahoe failed to do so.

For almost a half-century, the League to Save Lake Tahoe has been Tahoe's most responsible environmental watchdog. They have faithfully lived up to their mission statement: "The LTSLT is dedicated to protecting and restoring the environmental health, sustainability, and scenic beauty of the Lake Tahoe Basin. We focus on water quality and its clarity for the preservation of a pristine lake for future generations." The League's contributions in helping to preserve the lake's clarity and to protect its alpine environment are many.

Unarguably, the League is responsible for stopping the unregulated development that threatened the Basin before the Tahoe Regional Planning agency was formed—in fact, they are responsible for creating TRPA. Almost 60 years ago, a small group of Bay Area environmental activists campaigned to prevent the State of California from building a bridge across Emerald Bay. Emboldened by their success, the activists formed an organization that later morphed into the League to Save Lake Tahoe. They insightfully believed that the Basin's land use should be regulated by a common, highest-authority entity. To those forward-thinking LTSLT founders, TRPA can thank their existence.

That is why some longtime LTSLT members were surprised when in a December 2014 TRPA press release the agency took credit for halting "the run-away growth that Lake Tahoe was experiencing in the 1960s, ending plans for development of a city the size of San Francisco, and construction of a four-lane highway around the Lake with a bridge across Emerald Bay"—remarkable claims considering TRPA wasn't created until a decade after the League's founders had quashed the plans for the highway and bridge, and there was never a plan for a San Francisco-size city.

In 2013 I attended the League's inaugural two-day Eyes on the Lake training program. A half-day field trip showed a dozen aquatic plant neophytes, how to identify Eurasian watermilfoil and curlyleaf pondweed in stagnant swales along the lakeshore—it was instructive that both plants were found in thick patches within yards of the lake, but none had been able to take root in the lake's sandy bottom. However, as I drove away from Taylor Creek Beach, I couldn't help but think that the enthusiastic docents had missed the forest for the trees.

All of the 15 LTSLT staff listed on their website have a degree or experience in environmental science or related fields. Yet, Tahoe's foremost environmental advocacy group was focused on plants that have most likely established in all suitable Tahoe habitats and have almost no chance of infesting new areas in the lake, and they were ignoring the ecological devastation occurring under rubber mats a mile north of where they stood.

More than $30 million dollars has been spent to protect Tahoe from AIS and control those that are already living beneath its startlingly clear water—and not a single dollar of financial harm or any measurable environmental damage can be attributed to AIS at Lake Tahoe or any other Sierra lake. It's pretentious to think that mankind can control and improve on nature—at best, habitat disruptions and environmental abuses can be mitigated.

Outdoorsmen and conservationists know that nature's complexities aren't learned from lab experiments and textbooks, and until immersed in a natural habitat, ecological interactions and interdependencies are superficially understood. Backpackers, anglers, and sportsmen, such as Outdoors Hall of Fame inductee Tom Stienstra who has hiked 25,000 miles, camped countless nights under the stars, and writes the outdoors column for the *San Francisco Chronicle,* have a deeper understanding and more sensitive appreciation of nature's ways than do lab technicians.

Describing nature experiences lost on most self-declared environmental scientists, Stienstra wrote, "I've taken photos of fawns and foxes, snuck up on bears at 2 a.m., and stalked and [photographed] bucks at close range. Each encounter feels like being allowed entry into a secret world. I wonder if more people learned to love wildlife and how to find it, perhaps they would be more apt to protect it."

The cynicism I encountered from TRPA and Tahoe RCD managers charged with protecting Tahoe's environment and with oversight of environmental funds and of university administrators, who are tasked with overseeing the quality of their faculties' research and the science introduced by them, was astonishing. Professors, who received grants and funding totaling millions of dollars, were not only dismissive of science that disproved their theories but also egregiously distorted experiments and biased results to justify unfounded conclusions and recommendations and to cover prior errors and misstatements.

The process of forming Tahoe's AIS policy is neither inclusive nor transparent, and that is the fault of TRPA. But responsibility for the financially wasteful and environmentally destructive AIS programs that the agency administers primarily falls to the research professors who advise them. The criticisms of academics, researchers, and agency managers named in this book are harsh, as well as those of Tahoe RCD and the TRPA, but each declined to refute the allegations against them or disprove the aquatic science cited, and all cut off dialog after their errors and contradictions were presented. Together, the agencies and the academics who advise them have created the illusion that they are providing meaningful aquatic programs and protecting Lake Tahoe, while often doing more harm than good.

At Tahoe, aquatic environmental input is closed to all but the few who benefit from non-native species being classified as invasive, and tens of millions of dollars and valuable resources are wasted on protecting the lake from plants and animals that can't live there and controlling the few that have harmlessly carved an ecosystem niche. TRPA is charged with Tahoe's aquatic invasive species program oversight, and if their AIS management programs are necessary and effective, they can easily quiet the dissenters and resolve the conflicting viewpoints: All they have to do is have independent experts perform an AIS risk assessment.

A Final Note: From 2006 to 2014, Jeff Cowen served as TRPA spokesman and community liaison. When I was researching TRPA's programs, he was the most forthright of those contacted. His job was to pass on TRPA's public message, and he was often given the unenviable task of communicating misinformation. I sent him an early draft of *Tessie* for his review. Even though he no longer worked for TRPA and had only a few days to respond, he read the book and commented on six issues. They ranged from my assertion that most boaters don't believe paying up to $150 for a needless inspection and decontamination before launching in Lake Tahoe is minimally intrusive (the actual maximum cost with decontamination was $156) to my comment on Sudeep Chandra's statement: "It's not all about the mussel, if we make this just about mussels, we are in very deep trouble."

I had written: "It was all about the mussels for the first four years of the boat inspection program and is still mostly about the mussels." Cowen replied: "The purpose of inspections and the overall AIS program is to prevent the introduction of new aquatic species. Yes, this includes quagga and zebra mussels, but it also includes: New Zealand mud snail, hydrilla, Brazilian waterweed, parrot feather watermilfoil, warm-water fish species not present in Lake Tahoe, spiny water flea, and any other species with the potential to establish in the lake." Cowen concluded his email with: "I had hoped your book would convey to the reader the policy choice made by TRPA that involved erring on the side of caution. The agency has taken the position that Lake Tahoe is too important of an environmental and community resource to risk introduction of new invasive species like quagga mussels."

The issues couldn't be better framed. Indisputably, quagga and zebra mussels and New Zealand mud snails can't establish in Lake Tahoe. Warm-water fish densely populate only the Tahoe Keys, and there they do no harm. Spiny water flea would be a beneficial introduction, if they could establish. The aquatic plants Cowen lists can't prosper in Tahoe's climatic conditions or root on rocky substrate. And isn't it taking caution to an unreasonable extreme to decontaminate boats in case they might introduce an unknown species that may be able to establish in Lake Tahoe and cause harms like those from quagga mussels? Do Tahoe's environmental guardians really believe that a harmful species is going to pop up at Lake Tahoe before it's discovered in another North American lake? And until one does, isn't it irresponsible to inconvenience boaters and spend millions to attempt to keep benign and unknown species out of Lake Tahoe.

These are the issues that need to be put to rest. To read the studies that TRPA bases their aquatic invasive species policies upon and the articles that shaped public perception towards Tahoe's non-native aquatic species go to **SaveTessie.org**.

Writing about a Tahoe agency whose board he believes is ineffective, Ed Gurowitz wrote in June 2015: "Public disgust with the dysfunctional [board] is growing. Nothing will happen, though, unless that disgust translates into some sort of communication and action on the part of the residents of the district. ... So what is to be done? Well, one thing is to make sure that the conversation cannot be marginalized—as long as the voices of outrage are few, they can be ignored."

Gurowitz could just as easily have been writing about the TRPA, his former employer. Few are outraged over the economic waste and environmental harm of TRPA's AIS control programs, and Tahoe's boaters have been so thoroughly indoctrinated by TRPA's propaganda that they dutifully queue up to pay their fees for totally needless inspections and decontaminations.

If you are tired of being marginalized and believe that TRPA should conduct an independent AIS risk assessment, email **Joanne Marchetta (jmarchetta@trpa.org)** and **Casey Beyer (cbeyer@trpa.org)** to let them know of your disgust over their cynical disregard for the aquatic habitat and Tahoe environment that they are entrusted to protect.

Steve Urie

Works Cited

Akibso, Á., & Castro-Díez, P. (2012). Tolerance to air exposure of the New Zealand mudsnail as a prerequisite to survival in overland translocations. *NeoBiota*.

Bergquist, L. (2014, April 4). Study: Invasive zebra mussels improve water clarity in Great Lakes. *Detroit Free Press*.

Breining, G. (2009, November 19). Courting Controversy with a New View on Exotic Species. *Yale Environment 360*.

Brooks, D. (2012, July 2). Rainbow trout is an 'invasive' species that we try to spread - why? *The Telegraph*.

Bunker, D. (2013, May 10). Saga of the Quagga. *Moonshine Ink*.

Caldwell, T. J., & Chandra, S. (2012). Inventory of aquatic invasive species and water quality in lakes in the Lower Truckee River Region: 2012.

Chandra, S. (2009). Quagga Mussel Risk Assessment. University of Nevada, Reno.

Chandra, S. (2013, 2014). Science to assist policy decisions regarding the prevention of invasive species: testing the survival and growth of quagga mussel in Lake Tahoe, Annual Reports. U.S. Forest Service.

Chandra, S. (2013). Turning Tahoe's invasive crayfish into a consumable delicacy. Retrieved from tahoe.blogs.unr.edu

Chandra, S. & Wittmann, M. (2013). Development of a risk model to determine the expansion potential environmental impacts of Asian clams in Lake Tahoe.

Chapra, S. C. (2012). Long-term trends of Great Lakes major ion chemistry. *Journal of Great Lakes Research, 550*.

Claudi, R. & Prescott, K. (2011). Examination of Calcium and pH as Predictors of Dreissenid Mussel Survival in the California State Water Project. RNT Consulting.

Cobourn, J. & Segale, H. (2005, July 6) Divers remove invasive watermilfoil by hand. *Tahoe Daily Tribune*

Connelly, N.; O'Neill, C.; Knuth, B.; Brown, T. (2007). Economic Impacts of Zebra Mussels on Drinking Water Treatment and Electric Power Generation Facilities. Springer Science & Business Media, LLC

Cornell University Cooperative Extension. (2014) New York Invasive Species Information; The New York Invasive Species Clearinghouse Retrieved from Asian clam (Corbicula fluminea) http://nyis.info/?action=invasive_detail&id=52

CSGE. (2012). Quagga and Zebra Mussel Eradication and Control Workshop. California Sea Grant Extension and University of California Cooperative Extension. San Diego.

Davis, Mark; et al (2011, June 9). Don't judge species on their origins. *Nature*. Macmillan Publishers, Ltd.

Davis, Mark (2009). Invasion Biology, Oxford University Press, New York.

DeBolt, E. (2012, April 17). A Tale of Two Lakes: Lesson Learned from Lake Tahoe. *New York Legislative Gazette*, pg. 12.

Dietz, T. L. (1994). Osmoregulation in Dreissena polymorpha. *Biological Bulletin*, 187, 76-83.

Gamble, A.; Reuter, J.; Schladow, G.; Chandra, S. (2013) Emerald Bay control and management: stressors and mechanisms controlling Asian clam populations in Emerald Bay, Interim report on Lake Tahoe Restoration Projects.

Flanzraich, Annie. (2008, May 22). One mussel, one world of trouble. *North Lake Tahoe Bonanza.*

Fleeman, Michael. (2015, January 21). Lake Tahoe's tiny creatures dying off at dramatic rate: scientist. *Reuters*

Goldfarb, Ben. (2015, September 28). Can herbicides keep Tahoe blue? *High Country News.*

Goldfarb, Ben. (2015, September 2). Invasive crayfish in Oregon devastate native newts. *High Country News.*

Goode, Erica (2016, February 29). Invasive species aren't always unwanted. *NY Times.*

Hall, A. F. (2010, October 1). Invasives Bring Lake George and Lake Tahoe Scientists Together to Address Common Threats. *Lake George Mirror.*

Halverson, A. (2010). An Entirely Synthetic Fish: How Rainbow Trout Beguiled America and Overran the World. *Yale University Press.*

Hanson, Gretchen J.A., & et al (2013). Commonly Rare and Rarely Common: Comparing Population Abundance of Invasive and Native Aquatic Species. *PLOS ONE*

Herbst, D. B. (2008). Low specific conductivity limits growth and survival of the New Zealand mud snail from the Upper Owens River, California. *North American Naturalist.*

Howard, G. (2007, July 9). Bug Busters. *Sierra Sun.*

Ianniello, R.S. (2013) Effects of Environmental Variables on the Reproduction of Quagga Mussels in Lake Mead, NV/AZ UNLV/Theses/Dissertations/ProfessionalPapers/Paper184

Judson, H.F. (2004). The Great Betrayal: Fraud in Science. Harcourt, Inc., Orlando, FL

Kamerath, M.; Chandra, S.; & Allen, B. (2008). Distribution and impacts of warm water invasive fish in Lake Tahoe, USA. *REABIC Aquatic Invasions.*

Kolosovich, A., & et al (2011). Short-term survival and potential grazing effects of New Zealand mudsnail in an uninvaded Western Great Basin watershed. REABIC Aquatic Invasions.

Lovell, S. & Stone, A. (2005). The Economic Impacts of Aquatic Invasive Species: A Review of the Literature. National Center for Environmental Economics.

League to Save Lake Tahoe (LTSLT). (2013). Eyes on the Lake program materials.

Mackie, Gerald & Claudi, Renata. (2009). Monitoring and Control of Macrofouling Mollusks in Fresh Water Systems, Second Edition. CRC Press, Boca Raton, FL

Makeley, Michael; (2014) Saving Lake Tahoe: An Environmental History of a National Treasure, *University of Nevada Press.*

Marchetta, J. (2009, May 22). TRPA is working to keep Tahoe mussel free. *North Lake Tahoe Bonanza.*

Martin, G. (2013, February 5). Great Invaders: A new ecosystem is evolving in San Francisco Bay. *San Francisco Chronicle.*

McLaughlan, et al (2013). How complete is our knowledge of the ecosystem services impacts of Europe's top 10 invasive species? *Acta Oecologica.*

McLaughlan, C. & Aldridge, D.C. (2013) Cultivation of zebra mussels within their invaded range to improve water quality in reservoirs. *Water Research*

McMahon, R. & Bogan, A. (1991). Mollusca: Bivalvia; Ecology and Classification of North American Freshwater Vertebrates.

MDNR, M. D. (2013). Spiny water flea. Retrieved from AIS: dnr.state.mn.us/invasives/aquaticanimals/spinywaterflea

Menninger, H. (2013). The Asian clam, Corbicula fluminea: A brief review of the scientific literature. New York Invasive Species Research Institute.

Muskopf, S. (2007). The Effect of Beaver Dam Removal on Total Phosphorus Concentration in Taylor Creek and Wetland. Humboldt State University, Natural Resources.

Nalepa, T. F., & Schloesser, D. W. (2013). Quagga and Zebra Mussels: Biology, Impacts, and Control, *CRC Press.*

NISIC, N. I. (2013). Gateway to invasive species information; covering Federal, State, local, and international sources.

Retrieved from USDA Invasive Species Information: invasivespeciesinfo.gov/index.shtml

Ostroumov, S.A. (2005) Some aspects of water filtering activity of filter-feeders. Aquatic Biodiversity II, Developments in Hydrobiology.

Prescott, K. (2012). Examination of Water Quality in Clear Lake, CA for Dreissenid Mussel Suitability. RNT Consulting

Richards, G. (2004, April 30). Demise of the Lahontan Cutthroat Trout, Originally appeared in the *Sierra Sun*, Truckee-Donner Historical Society.

Richter, A. (2008). Pacific Northwest Aquatic Invasive Species Profile, U.S. Geological Survey

Rosa, I. C. (2012). Effects of Upper-Limit Water Temperatures on the Dispersal of the Asian Clam. *PLOS-One*.

Rosaen, et al (2012). The Costs of Aquatic Invasive Species Anderson Economic Group.

Rosamond, C. (2013, May 10). Harvest Corbicula, the Not So "Golden" Clam. *Moonshine Ink*.

Ruhmann, E. (2014). Survival, growth, and settlement of Dreissena rostriformis bugensis veligers in low and high calcium waters, UNLV University Libraries. Retrieved from: digitalscholarship.unlv.edu/cgi/viewcontent.cgi?article=3215&context=thesesdissertations

Sahagun, L. (2013, September 2). Poisoning a Sierra stream to save the world's rarest trout. *Los Angeles Times*.

Sax, Dov F., et al (2005). Species Invasions: Insights Into Ecology, Evolution, BioGeography. Sinauer Associates, Sunderland, Massachusetts.

Schladow, G. (2008, December 15). California Colloquium on Water—Schladow Retrieved from UC Berkeley Events: youtube.com/watch?v=AQDqDKGohY4&feature=related

Schladow, G. (2011). Lake Tahoe AIS Public Forum.

youtube.com/watch?v=wiscdjG1Cz4

ScienceDaily. (2011, June). Mystery Solved: Scientists Discover How 'Didymo' Algae Bloom in Pristine Waters With Few Nutrients. *Science News*.

Schreiber, E.S.G., et al. (2002). Distribution of an alien aquatic snail in relation to flow variability, human activities, and water quality. *Freshwater Biology* 48: 951-961.

Scott, E.B. (1957) The Saga of Lake Tahoe. Sierra-Tahoe Publishing Co., Chrystal Bay, Nevada.

Shyama, P. (2005, January 24). Global Invasive Species Database (Asian clams). Retrieved from AIS Specialist Group: issg.org/database/species/ecology.asp?

Stienstra, T. (2012). California Fishing: The Complete Guide to Fishing on Lakes, Streams, Rivers, and the Coast, Moon Outdoors

Stienstra, T. (2014, January 27). Fish in a barrel? Not how nature rolls. *The San Francisco Chronicle*.

Strayer, D.L., et al. (1999) Transformation of ecosystems by bivalves. *Bioscience* 49: 19-27

TERC. (2012). Tahoe: State of the Lake Report 2011. UC Davis.

TERC. (2013). Tahoe: State of the Lake Report 2012. UC Davis.

TERC. (2014). Tahoe: State of the Lake Report 2013. UC Davis.

Therriault, T. (2010). Risk assessment for New Zealand mud snail in Canada. Fisheries and Oceans Canada.

TRPA. (2009). Lake Tahoe Region AIS Management Plan.

TRPA. (2014). Lake Tahoe Region AIS Management Plan - Update

TRPA. (2009, April 21). Protect the Lake This Summer. *Tahoe Daily Tribune*.

UPI. (2009, August 19). Tiny Asian clams invade Lake Tahoe. *UPI Science*.

USACE, (2013). Invasive Species Information System. Retrieved from Engineer Research and Development Center

Vander Zandem, M. J. (2007). Surveillance and Control of Aquatic Invasive Species in the Great Lakes. Madison: University of Wisconsin.

Walker, B. (2010). Big Price -- Little Benefit: Proposed Locks on the Upper Mississippi and Illinois Rivers Are Not Economically Viable. Nicollet Island Coalition.

Watson, B. (2010). 10 Invasive Species That Cost the U.S. a Bundle. *Daily Finance.*

Weisser, P. (2012) SWP reservoir Boat Inspections Intensify in DWR's Invasive Mussel Control Program. California *DWR News.*

Whittier, T. R., et al. (2008). A calcium-based invasion risk assessment for zebra and quagga mussels *(Dreissena Spp)*. *Frontiers in Ecology and the Environment.*

Wittmann, M.; Chandra, S.; Reuter, J.; Caires, A.; Schladow, G.; Denton, M. (2012) Harvesting an invasive bivalve in a large natural lake: species recovery and impacts on native benthic macroinvertebrate community in Lake Tahoe, USA

Wittmann, M.; Chandra, S.; Schladow, G.; & Reuter, J. (2013). Natural and Human Limitations to Asian Clam Distribution and Recolonization—Factors that Impact the Management and Control in Lake Tahoe.

Wittmann, M.; and Chandra, S. (2015). Implementation Plan for the Control of Aquatic Invasive Species within Lake Tahoe.

Wittmann, M.; Reuter, J.; Schladow, G.; Hackley, S.; Allen, B.; Chandra, S.; et al. (2008). Asian clam of Lake Tahoe: Preliminary findings in support of a management plan.

Zimmer, C. (2008, September 8). Friendly Invaders. *NY Times.*

Index

100th Meridian Initiative, 15
2009 *Lake Tahoe Region AIS Management Plan*, 128, 169, 175
2011 AIS Forum, 100
2013 AIS Forum, 220
2014 AIS Forum, 211
2014 *Lake Tahoe Region AIS Management Plan*, 178
2015 AIS Forum, 22, 110, 130
Acharya, Kumud, 60, 61, 69, 153
Ajari, Bruce, 166
Allen, Brant, 41, 42
Ames, Laurel, 183
Anderson Economic Group, 176, 178
Anderson, Lars, 130, 132
Angle, Sharron, 181
ANS Task Force (ANSTF), 9, 20, 167, 176, 178
Archarya, Kumud, 71
Atwell, Lisa, 149
Baggerly, Russ, 89
Bauer, Peter, 142
Beyer, Casey, 68, 69, 84, 112, 200, 229
Boosman, Kristy, 27
Bresnick, Mara, 33
Britton, David, 39, 40
Brown University, 17
Brown, James, 18
Brown, Jerry, 183
Bunker, David, 52, 53, 201
Caldwell, Timothy, 153
California Colloquium on Water, 43, 216
California Department of Boating and Waterways, 208
California Department of Fish and Wildlife, 39, 152, 195, 196
California Department of Food and Agriculture, 15, 134
California Department of Parks and Recreation, 101
California Department of Water Resources (CDWR), 62, 71, 152, 164, 208, 217
California Quagga and Zebra Mussel Interagency Team, 208
California State Lands Commission, 121
California Tahoe Conservancy, 120, 170
Carson, Kit, 28
Cartwright, Nicole, 22, 72, 93, 191, 212, 213
Chandra, Sudeep, 33, 34, 38, 50–86, 93, 94, 102, 117–19, 124–27, 139, 140, 154, 162–64, 167, 188, 191, 197–204, 206, 210–11, 221, 224, 228
Chilton, Steve, 191
Claudi, Renata, 50, 62, 184, 209
Clinton, Bill, 30, 115
Collins, Darcie Goodman, 157, 158, 159
Cornell University, 45, 100, 176, 188
Cousteau, Jacques, 11, 12
Cowen, Jeff, 108–10, 185, 186, 201, 220, 227, 228
Crimmens, Teresa, 151–54, 160, 163, 164, 166, 167, 220
Crookshanks, Chris, 205
Dahlgren, Kristen, 117
Darrin Fresh Water Institute, 53, 144
Davis, Clinton, 149
Davis, Mark, 23, 24, 25, 47, 195, 197
DeBolt, Emily, 137
Desert Research Institute (DRI), 60, 68, 98
Dunbar, Madonna, 190
Earthjustice, 183
Ecological Society of America, 48
EPA Monitoring and Assessment Program, 48

EPA Wadeable Streams Assessment, 48
Escobar, Pablo, 59
Feinstein, Dianne, 1
Franklin, Jerry, 175
Fremont, John, 28, 179
Fund for Lake George, 142, 145
Geist, Willie, 117
Goldfarb, Ben, 131
Goldman, Charles, 29, 30, 64, 188, 202, 203
Gore, Al, 30
Grant, Jim, 11
Gurowitz, Ed, 181–84, 223, 229
Hackley, Scott, 42
Holan, Lisa, 165
Holzer, Katie, 188
Jardine, Don, 196
Julienne, Jason, 152, 153
Kennedy, John F., 211
Kolosovich, Alexander, 149
Lahontan Regional Water Quality Control Board, 121, 130, 132, 139, 190, 191
Lake Champlain AIS Management Program, 144
Lake George Asian Clam Rapid Response Task Force, 143
Lake George Association, 137, 143
Lake George Park Commission, 142, 146
Lake Tahoe AIS Coordination Committee, 222
Lake Tahoe Aquatic Invasive Species Program, 109
Lake Tahoe Asian Clam Working Group, 1, 109, 185
Larson, John, 130
League to Save Lake Tahoe, 11, 157, 159, 218, 224, 225
Lentz. Dave, 195
Lind, Rick, 130
Lotshaw, Tom, 111

Louisiana State University (LSU), 55
Macalester College, 23, 47, 195
Mackie, Gerald, 50, 55, 62, 184
Marchetta, Joanne, 91, 94, 95, 131, 133, 182, 213, 214, 229
McLaughlan, Claire, 75
Meder, Martin, 173
Modley, Meg, 144
Mountain Area Preservation Foundation, 165
Muir, John, 98
Mussel Mast'R, 91
Nagy, Lester, 29
National Center for Environmental Economics (NCEE), 176, 178
National Science Foundation, 156
Negrete, Adilene, 133
Nevada Division of Wildlife, 203, 205
Nierzwicki-Bauer, Sandra, 53, 144
Novitsky, Chris, 145
Oliver, Dennis, 42, 44
Olsen, Dan, 151
Oregon State University, 13, 48, 76
Ostroumov, S.A., 60
Palma, Juan, 182
Penrose, Ron, 152, 160, 161
Query, Toby, 197
Regan, Julie, 14, 15, 215
Reuter, John, 44, 102, 137, 166, 191
RNT Consulting, 62, 152, 153, 208
Robbins, James, 12
Roberts, Jason, 39
Roefer, Peggy, 80, 81, 161
Roker, Al, 117
Rosamond, Chris, 98
Rudroff, Savannah, 134
Saito, Laurel, 149
Sax, Dov, 17
Schladow, Geoffrey, 43, 97, 99–108, 137, 143, 175, 187, 188, 191, 201, 211, 216, 219, 224
Shaw, Dan, 122

Sierra Club, Toiyabe Chapter, 183
Sierra Ecosystem Associates, 130, 132
Sierra Wildlife Coalition, 23
Simon, Judith Michaels, 131
Singlaub, John, 39, 182
Siy, Eric, 145
South Tahoe Public Utility District, 131
Southern Nevada Water Authority (SNWA), 80, 81, 160, 161
St. Denis, Gene, 203
Stienstra, Tom, 127, 226
Stone, Patrick, 122, 205
Tahoe Environmental Improvement Program (EIP), 4, 51, 120, 186, 217
Tahoe Environmental Research Center (TERC), 41–44, 48, 51, 99, 102, 103, 111–15, 117, 121, 127, 147, 184, 186, 201, 211, 220
Tahoe Keys Property Owners Association, 130, 190, 212
Tahoe Lobster Co., 203
Tahoe Regional Planning Agency (TRPA), 13–16, 22, 47, 50, 52, 80–83, 87–95, 100–104, 111–20, 123, 126, 135, 140, 147, 151, 157, 159, 168, 170, 172, 175, 212–17, 223–28
Tahoe Resource Conservation District (Tahoe RCD), 22, 82, 83, 93, 104, 108, 116, 124, 147, 151, 154, 155, 158–61, 166, 167, 178, 199, 205, 206, 211, 213, 220, 227
Tahoe Science Consortium, 78
Tahoe Water Suppliers Association, 190
Thayer, Ted, 40, 83, 87, 90, 91, 92, 108, 126, 139, 145, 170, 186, 191, 201, 219, 224
Thompson, Ken, 18, 23
TKPOA Water Company, 130
Trout Unlimited, 211
TRPA Board of Governors, 14, 33, 43, 54, 184

Truckee Meadows Water Authority (TMWA), 152, 160, 161
Truckee River Fund, 160
Truckee River Watershed Council (TRWC), 151, 163, 166
Truckee Town Council, 154, 155
Trumbo, Joel, 131
Twain, Mark, 218
U.S Fish and Wildlife Service (USFWS), 15
U.S. Army Corps of Engineers (USACE), 14, 90, 170
U.S. Bureau of Land Management, 51, 162
U.S. Bureau of Reclamation, 82, 152, 153, 168, 184
U.S. Environmental Protection Agency (EPA), 48, 153, 190, 201
U.S. Fish and Wildlife Service (USFWS), 39, 196
U.S. Forest Service, 22, 51, 61, 65, 66, 82, 102, 103, 118, 124, 141, 185, 196, 220
U.S. Geological Survey, 139
U.S. National Park Service, 206
UC Division of Agriculture and Natural Resources, 197
University of California, Berkeley, 43
University of California, Davis (UC Davis), 32, 41, 101, 137, 202
University of Michigan, 29
University of Nevada, Las Vegas (UNLV), 69, 76
University of Nevada, Reno (UNR), 2, 32, 43, 44, 51, 93, 117, 121, 123, 126, 137, 147, 151, 184, 203
University of New Mexico, 18
University of Wisconsin, 21, 111
University of Wisconsin-Oshkosh, 173
UNR Aquatic Ecosystems Laboratory, 33, 118
Vender Zandem, M. Jake, 21

Vermont Fish and Wildlife, 156
von Seggern, David, 183
Wake Worx, 91
Watanabe, Shohei, 138
Western Regional Panel on Aquatic Nuisance Species, 9, 21, 73
Whittier, Thomas R., 48
Wick, Dave, 142, 144
Willoughby, David, 152, 153, 164
Wittmann, Marion, 43, 48, 97, 98, 102, 105, 113, 131, 139, 140, 223
Zabaglo, Dennis, 72, 107, 185, 212, 213, 220

www.ingramcontent.com/pod-product-compliance
Lightning Source LLC
Chambersburg PA
CBHW051638170526
45167CB00001B/235